International Conference on Building Materials and Materials Engineering (ICBMM 2017)

IOP Conference Series: Materials Science and Engineering
Volume 264

Lyon, France
21 – 23 September 2017

ISBN: 978-1-5108-5310-2
ISSN: 1757-8981

Printed from e-media with permission by:

Curran Associates, Inc.
57 Morehouse Lane
Red Hook, NY 12571

Some format issues inherent in the e-media version may also appear in this print version.

Printed by Curran Associates, Inc. (2018)

For permission requests, please contact the Institute of Physics
at the address below.

Institute of Physics
Dirac House, Temple Back
Bristol BS1 6BE UK

Phone: 44 1 17 929 7481
Fax: 44 1 17 920 0979

techtracking@iop.org

Additional copies of this publication are available from:

Curran Associates, Inc.
57 Morehouse Lane
Red Hook, NY 12571 USA
Phone: 845-758-0400
Fax: 845-758-2633
Email: curran@proceedings.com
Web: www.proceedings.com

Table of contents

Volume 264

2017 International Conference on Building Materials and Materials Engineering (ICBMM 2017)

21–23 September 2017, Lyon, France

Accepted papers received: 30 October 2017
Published online: 29 November 2017

Preface

Papers

IOP Conf. Series: Materials Science and Engineering **264** (2017) 011001 doi:10.1088/1757-899X/264/1/011001

PREFACE

It is our great pleasure to welcome you to 2017 International Conference on Building Materials and Materials Engineering (ICBMM 2017) which was held in Lyon, France during September 21-23, 2017. ICBMM 2017 is dedicated to issues related to Building Materials and Materials Engineering.

The major goal and feature of the conference is to bring academic scientists, engineers, industry researchers together to exchange and share their experiences and research results, and discuss the practical challenges encountered and the solutions adopted. Professors from France and Portugal are invited to deliver keynote speeches regarding latest information in their respective expertise areas. It will be a golden opportunity for the students, researchers and engineers to interact with the experts and specialists to get their advice or consultation on technical matters, sales and marketing strategies.

This proceedings present a selection from papers submitted to the conference from universities, research institutes and industries. All of the papers were subjected to peer-review by conference committee members and international reviewers. The papers selected depended on their quality and their relevancy to the conference. The volume tends to present to the readers the recent advances in the field of Materials Science and Engineering, Materials Properties, Measuring Methods and Applications, Civil and Structural Engineering, Architecture and Urban Planning etc.

We would like to thank all the authors who have contributed to this volume and also to the organizing committee, reviewers, speakers, chairpersons, sponsors and all the conference participants for their support to ICBMM 2017.

Prof. WONG Kwai Kwan
CNRS (National Center of Scientific Research); University of Lyon/ENTPE, France
October 19, 2017

IOP Conf. Series: Materials Science and Engineering **264** (2017) 011002 doi:10.1088/1757-899X/264/1/011002

Organizing Committees

Conference Chairs

Prof. Carlos Chastre, NOVA University of Lisbon, Portugal

Assoc. Prof. Paulo Mendonça, University of Minho, Portugal

Technical Program Committee Chairs

Prof. Giuseppe Bonifazi, University of Rome, Italy

Assoc. Prof. Mohammad Arif Kamal, Aligarh Muslim University, India

Organizing Chair

Ms. Yang Dan, Hong Kong Chemical, Biological & Environmental Engineering Society, Hong Kong

Publication Chair

Ms. Tammy Luo, Chemistry and Materials Society, Hong Kong Chemical, Biological & Environmental Engineering Society, Hong Kong (CMS-HKCBEES)

Technical Program Committee

Prof. Francisco M. Marquez Linares, University of Turabo, USA

Prof. Carlos Chastre, NOVA University of Lisbon, Portugal

Prof. S. Joseph Antony, University of Leeds, UK

Prof. Thomas Neitzert, Auckland University of Technology, New Zealand

Prof. Giuseppe Bonifazi, University of Rome, Italy

Prof. Alessandro Pegoretti, University of Trento, Italy

Prof. Hamit Yurtseven, Middle East Technical University, Turkey

Prof. Alexandre Maitre, Universite de Limoges, France

Prof. Veturia Chiroiu, The Romanian Academy, Romania

Prof. Nuno Dinis Cortiços, University of Lisboa, Portugal

Prof. Tadahiro Wada, Nara National College of Technology, Japan

Prof. Mosaad A. Abdel-Wahhab, National Research Center, Egypt

Assoc. Prof. Hessam Ghasemnejad, Cranfield University, UK

Assoc. Prof. François Barthelat, McGill University, Canada

Assoc. Prof. Khairunisa Muthusamy, Universiti Malaysia Pahang, Malaysia

Dr. Mohammad Reza Allazadeh, University of Strathclyde, UK

Dr. Hui Li, Washington State University, USA

Dr. Yingtao Liu, University of Oklahoma, Norman, USA

Dr. Xiaosong Huang, General Motors Global R&D, USA

Dr. Hugo Biscaia, Nova University of Lisbon, Portugal

IOP Conf. Series: Materials Science and Engineering **264** (2017) 011003 doi:10.1088/1757-899X/264/1/011003

Peer review statement

All papers published in this volume of *IOP Conference Series: Materials Science and Engineering* have been peer reviewed through processes administered by the proceedings Editors. Reviews were conducted by expert referees to the professional and scientific standards expected of a proceedings journal published by IOP Publishing.

IOP Conf. Series: Materials Science and Engineering **264** (2017) 012001 doi:10.1088/1757-899X/264/1/012001

Study on strength estimation of soil cement used in the embedded pile method by electrical resistivity measurement

Y Mochida[1], Y Sakurai[2], H Indra[2], A L Karimi[2]

[1] Professor of Graduate School of Engineering,
Ritsumeikan University, Japan
[2] Master Degree Student of Graduate School of Engineering,
Ritsumeikan University, Japan

Abstract. Problems caused by poor quality control and quality assurance of the pre-boring embedded pile construction, such as on domestic apartment house is still occurring nowadays. An adequate consideration for invisible risks inside or below the ground is important in pile foundation construction therefore the demand for advanced and reliable quality assurance is increase in the future. In this research, to understand the quality of the construction at early stage, the compressive strength of cement-soil mixture of pile construction after 28 days is estimated using electrical resistivity value of the mixture. More accurate measurement for electrical resistivity value is conducted by inserting the electrodes without using potassium chloride solution as a catalyst. The result showed that there is a certain tendency in the electric resistivity value at the early age regarding to the type of soil (sand, clay) mixed in. The most accurate estimation was achieved from the electric resistivity value at the first day and several days onwards, and from the compressive strength after 3 days.

1. Introduction

The purpose of this research is to study the utilization of electrical resistivity measurement to predict compressive strength and to perform quality control of cement slurry used in bored precast pile construction technique [1][2][4][5]. This paper consists report and recommendation for electrical resistivity measurement technique [3], quantification and estimation technique of electrical resistivity, compressive strength estimation formula based on electrical resistivity value and quality control method using strength prediction of cement slurry through indoor/laboratory and field examination [3][4][5].

2. Laboratory experiment of soil cement mixture

The electrical resistivity measurement methods were decided prior to laboratory examination. The following three patterns of electrical resistivity measurement methods were performed at the same time and suitable method is studied.

1. At 28 days from pouring without removing cell and no pore water.
2. Only measure bleeding water at early stage by making hole with diameter same as cell in specimens. After that stage, 0.1% KCl solution is used as pore water.
3. Using KCl solution with the same electrical resistivity value as specimen right after poured into cell. After that stage, the same procedure as number 2.

Considering actual soil configuration, cement slurry only, cement with sand, cement with clay and cement with sand and clay, three variants of mixture is used for laboratory examination. Electrical conductivity meter and schematic view of measurement are shown in Figure 1.

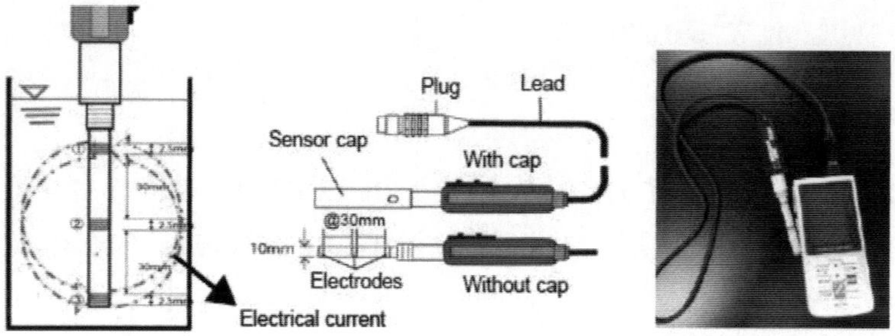

Figure 1. Electrical conductivity meter and measurement

2.1 Laboratory examination mixture using cement slurry

Electrical resistivity value change with time shown by Figure 2 gives information that electrical resistivity value is tends to be the same regardless of mixture and cement amount. For mixture curing age of one day, electrical resistivity value is greatly affected by pore water for method number 1 and number 2, therefore, method number 1 is recommended for electrical resistivity measurement. In addition, temperature change also influences electrical resistivity value.

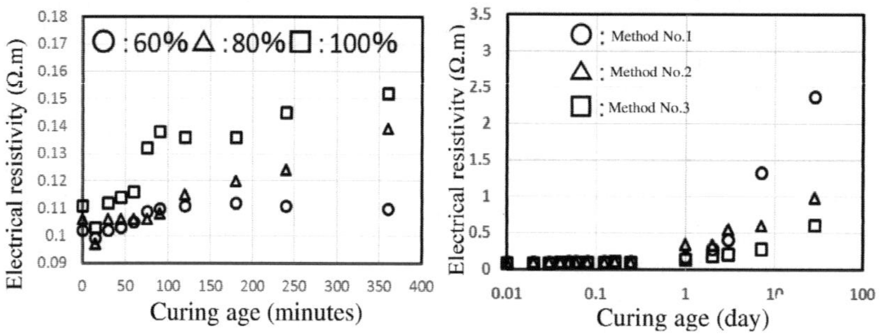

Figure 2. Electrical resistivity value change with time

Relationship between compression strength and electrical resistivity value measured by method number 1 as shown by Figure 3 resulted to conclusion that relation between qu_{28}/ρ_{28} and qu_3/ρ_3 was more correlative than $\Delta qu/\Delta\rho$ and qu_3/ρ_3. In other hand, measurement methods number 2 and 3 also had similar result, therefore, compression strength estimation formula using relationship between qu_{28}/ρ_{28} and qu_3/ρ_3 is recommended. Here, qu is compressive strength, ρ is electrical resistivity and subscript number indicates specimen curing age in day(s) unit.

Method 1 : $qu_{28}=\{0.1058.(qu_3/\rho_3)+8.8964\}\rho_{28}$ (1)

Method 2 : $qu_{28}=\{3.3823.(qu_3/\rho_3)-63.551\}\rho_{28}$ (2)

Method 3 : $qu_{28}=\{0.5807.(qu_3/\rho_3)+9.3560\}\rho_{28}$ (3)

Since electrical resistivity value tends to change regarding to time (curing age), compressive strength can be estimated using ρ value from early curing age until reached final curing age (28 days). Electrical resistivity changes due curing age is observed to decide the suitable time for estimating the compressive strength as shown by Figure 4. Relationship between ρ_3 and ρ_{28} recommends equation 4, equation 5 and equation 6 for estimating ρ value. In addition, considering ρ value at early age and onwards due to pore water existence, relationship between ρ_0 and ρ_3 is used for measurement method 1 and ρ_1 vs ρ_3 for method 2 and 3 recommends equation 7 ~ equation 9 for estimating ρ value. Therefore, equation 4 ~ equation 9 is used for estimating electrical resistivity value (ρ).

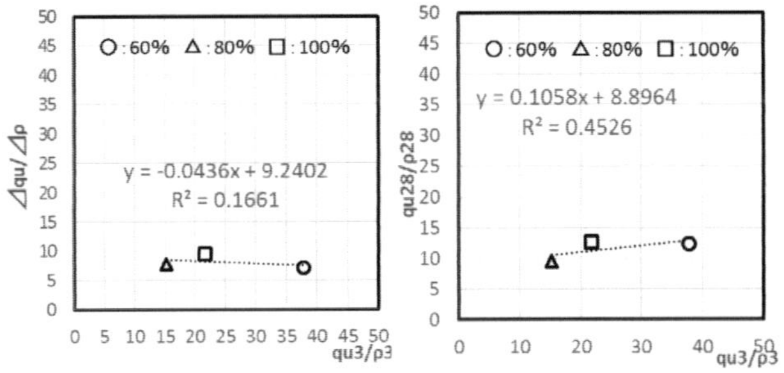

Figure 3. Relationship between $\Delta qu/\Delta\rho$ and qu_3/ρ_3 and qu_{28}/ρ_{28} and qu_3/ρ_3

$$\text{Method 1}: \rho_{28} = 7.8687 \cdot \rho_3^{1.5779} \quad (4)$$
$$\text{Method 2}: \rho_{28} = 0.8950 \cdot \rho_3^{0.153} \quad (5)$$
$$\text{Method 3}: \rho_{28} = 3.5787 \cdot \rho_3^{1.1367} \quad (6)$$
$$\text{Method 1}: \rho_3 = 0.0015 \cdot \rho_0^{-2.483} \quad (7)$$
$$\text{Method 2}: \rho_3 = 1.8722 \cdot \rho_1^{1.1642} \quad (8)$$
$$\text{Method 3}: \rho_3 = 4.0211 \cdot \rho_1^{1.5936} \quad (9)$$

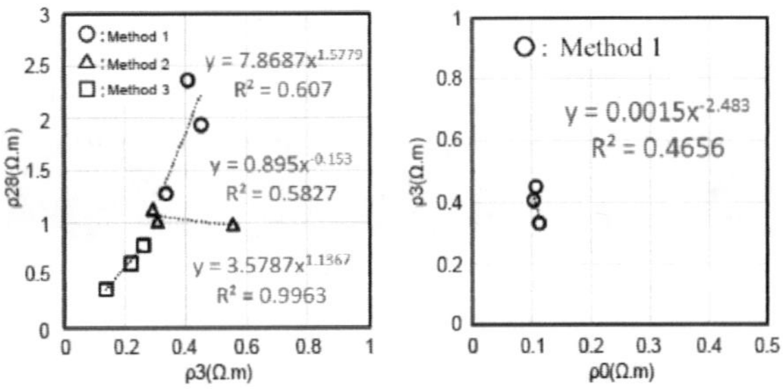

Figure 4. Relationship between ρ_3 and ρ_{28} and ρ_0 and ρ_3

Subsequently, estimation formulation for compressive strength prediction was also attempted. Equation 10 is formulation recommended by Japan Society of Civil Engineering (JSCE) [6], is used and applied for laboratory examination in this research. Where α and β is material constants and t is specimen's curing age. Calculation of performed by linear interpolation of $q_u(28)/q_u(t)$ vs t recommended by JSCE (Figure 5) and listed in Table 1.

Table 1. Laboratory experiment mixture condition

Sample condition (W/C)	α	β
60%	0.84	4.86
80%	0.68	9.53
100%	0.79	6.08
Average	0.77	6.6233

$$qu(\mathrm{t}) = qu(28) \times \frac{\mathrm{t}}{\alpha \times t + \beta} \quad (10)$$

Estimation of compressive strength using equation 1 to equation 9 and laboratory compressive strength test are attempted and the result is compared as shown by Figure 6. Estimated value and measured value are nearly consistent for measurement method 1. Estimated value is larger than measured value for method 2 while Measured value is larger than estimated value for method 3. Hence, method 1 is possible and recommended for estimating electrical resistivity value.

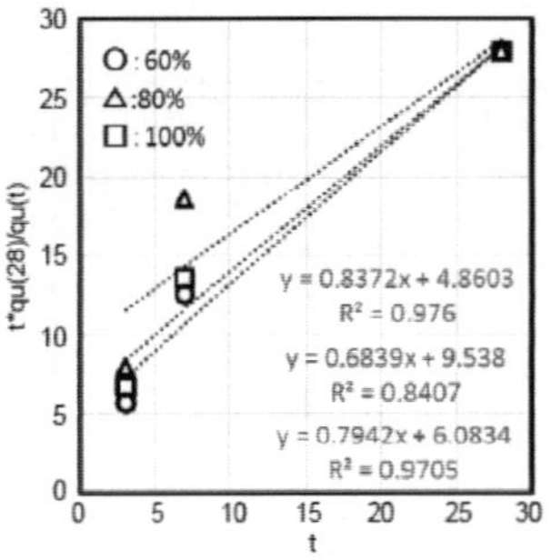

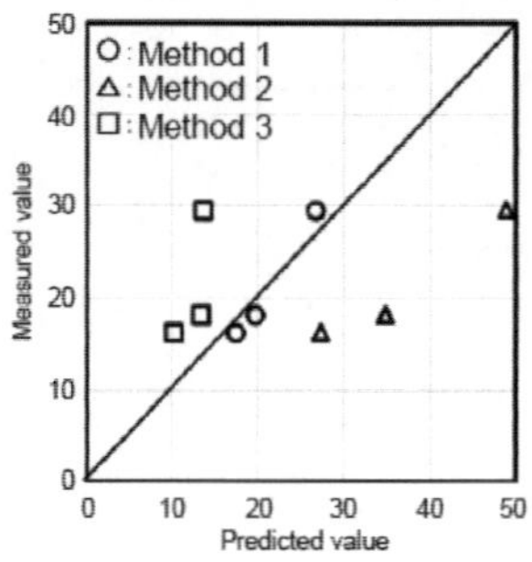

Figure 5. Relationship of t and $t.q_{u28}/qu_t$ (left) and **Figure 6.** Comparison of measured vs predicted value of compressive strength (right)

2.2 Laboratory examination mixture using sandy soil and clayey soil

Laboratory examination for sandy and clayey soil assumption are performed and the procedures are shown by Figure 7 and mixture condition listed in table 2. Electrical resistivity value changes by material age (time) for sandy soil is different from cement only and cement with clay soil due to precipitation influence. For clay soil, specimen is expanding and cracks occurs during curing process causing variation in electrical resistivity value. Furthermore, for clay soil, the viscosity of the soil cement slurry was very high therefore the compressive strength test could not performed as shown by Figure 8.

Figure 7. Soil cement mixing procedure

Table 2. Sandy soil assumed mixture condition

W/C ratio (%)	Mud water ratio	Cement milk : mud water		Replacement soil percentage (%)
60	1.5	1:3	A	30
60	1.5	1:5	B	20
60	1.5	1:10	C	10
100	1.5	1:2	D	50
100	1.5	1:2.5	E	40
100	1.5	1:2	F	30

Compressive strength estimation for sandy soil specimen are also conducted using the same procedure as cement slurry only specimen and recommended formulation, equation 11 ~ equation 19 and equation 10, and comparison between measured and estimated value is shown by Figure 9. As a result, for every measurement method, compressive strength prediction is possible even when sand is mixed. There was specimen with large error of compressive strength value, however, bleeding water effect during curing process might cause this phenomenon.

Method 1 : $qu_{28}=\{0.2514 \cdot (qu_3/\rho_3)+0.6748\} \cdot \rho_{28}$ (11)

Method 2 : $qu_{28}=\{0.451 \cdot (qu_3/\rho_3)+0.5002\} \cdot \rho_{28}$ (12)

Method 3 : $qu_{28}=\{0.9578 \cdot (qu_3/\rho_3)+2.9513\} \cdot \rho_{28}$ (13)

Method 1 : $\rho_{28}= 0.2807 \cdot \rho_3^{-2.071}$ (14)

Method 2 : $\rho_{28}= 0.7269 \cdot \rho_3^{-0.2769}$ (15)

Method 3 : $\rho_{28}= 0.9634 \cdot \rho_3^{0.572}$ (16)

Method 1 : $\rho_3= 0.5048 \cdot \rho_0^{0.2804}$ (17)

Method 2 : $\rho_3=0.0946 \cdot \rho_1^{-0.599}$ (18)

Method 3 : $\rho_3=0.7643 \cdot \rho_1^{0.7377}$ (19)

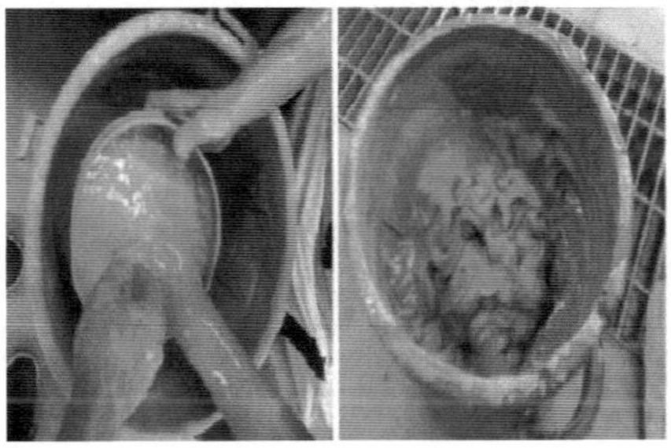

Figure 8. Experiment on clay soil

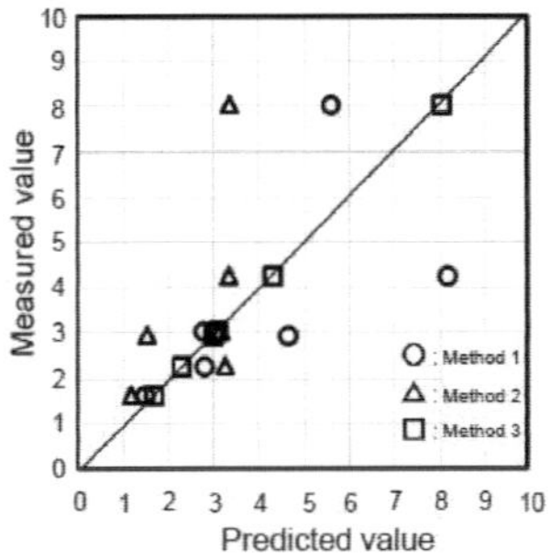

Figure 9. Predicted vs measured value

2.3 Laboratory mix examination assuming actual ground

Laboratory mixing experiment assuming actual soil condition was performed using sand and clay mixed together with mixing conditions listed in table 3. The experiment results lead to inconsistent change of electrical resistivity value with time along with large variation. Similar method to predict compressive strength of the specimen using relationship between qu_{28}/ρ_{28} and qu_3/ρ_3 also applied in this procedure and came to results that equations 20~28 are recommended for compressive strength prediction. Comparison between predicted value and measured value shown by Figure 10.

Method 1 : $qu28=\{0.2662(qu3/\rho3)+1.9764\}\rho28$(25)

Method 2 : $qu28=\{4.1042(qu3/\rho3)-81.007\}\rho28$(26)

Method 3 : $qu28=\{2.8788(qu3/\rho3)-51.029\}\rho28$(27)

Method 1 : $\rho28 = 3.2724\rho3^{0.6215}$(28)

Method 2 : $\rho28 = 66.139\rho3^{3.2056}$(29)

Method 3 : $\rho28 = 278.01\rho3^{4.227}$(30)

Method 1 : $\rho3 = 0.5048\rho0^{0.2804}$(31)

Method 2 : $\rho3 = 0.7679\,\rho1^{0.7792}$(32)

Method 3 : $\rho3 = 0.4656\,\rho1^{0.4635}$(33)

Table 3. Sand with clay soil assumed mixture condition

W/C ratio (%)	Mud water amount ratio	Cement milk : mud water		Sand : clay	Replacement soil percentage (%)
60	1.6	1:1	G	3:7	100
60	1.6	1:1	H	6:4	100
60	1.6	1:1	I	8:2	100

Measurement method 1 and 2 gives prediction with high accuracy, however not with method 3. It is understood that the mixture of sand and clay has moderate density and reduce bleeding process, therefore increase accuracy in measuring electrical resistivity and improves accuracy of prediction formulation.

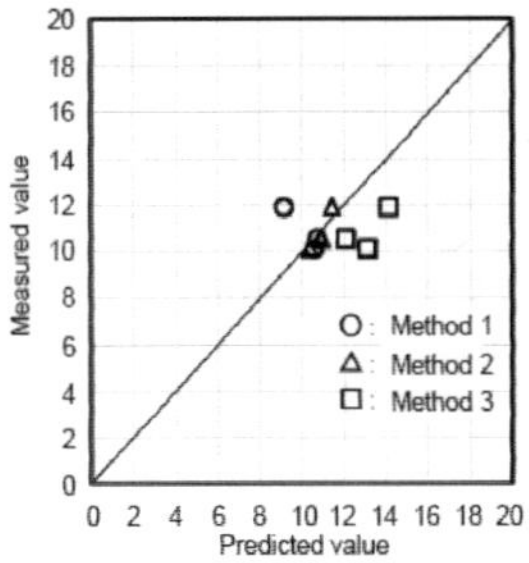

Figure 10. Predicted vs measured value for sand with clay soil assumed specimen

IOP Conf. Series: Materials Science and Engineering **264** (2017) 012001 doi:10.1088/1757-899X/264/1/012001

3. Outdoor experiment

3.1 Examination of pile hole at the construction site.

Bored pile construction site in Arahama, Kashiwa city of Niigata prefecture is selected for outdoor experiment. The specifications were bored pile hole with diameter of 530mm, depth of 20m and with W/C ratio of 60%. Collection of unsolidified sample was carried out from overflow and mortar plant and the universal water sampler shown in Figure 11 and was conducted from depth of 2.5m, 5.0m, 10m, and the specific resistance value on the ground are Measured. Soil until depth of 15m was sandy soil. At the same time, electrical resistivity value near depth of 5.0 m was also measured using custom made electrical conductivity meter with cord length of 30 m.

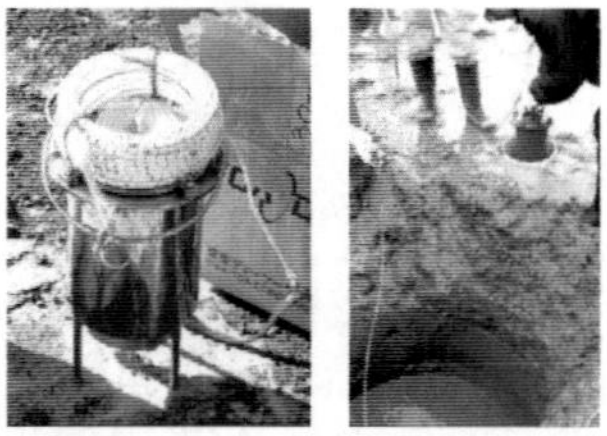

Figure 11. Tools and sample collection

Changes of electrical resistivity value was rapid as shown by Figure 12. Is it believed that summer season condition caused the acceleration of specimen curing process. Figure 12 also shows that the temperature rise is extremely low inside the hole, the influence of the outside air is small at the depth of 5.0m and the influence on the quality by the temperature is not significant.

The result of compressive strength value using prediction equation is shown as listed in table 4. An Accurate strength prediction performed by sample collected from inside the bored hole. In addition, it is possible to predict the compressive strength with an error of about 1.0 N / mm² for safety consideration. However, for sample collected from the plant gives more variation as effect of temperature change is believed to cause this large margin of error.

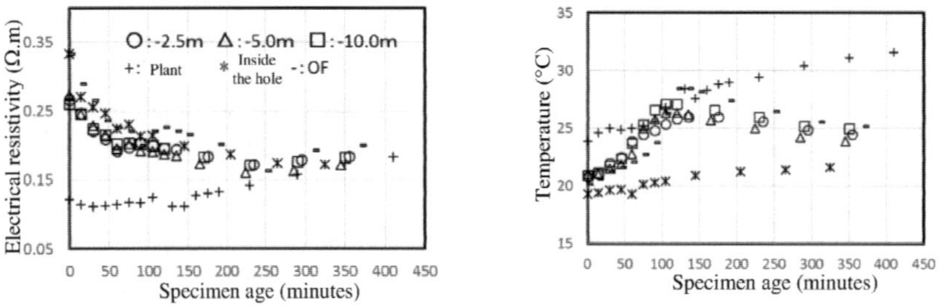

Figure 12. Electrical resistivity value and temperature change

Table 4. Compressive strength of specimen at each collection location

	Plant	Depth = 0m	Depth = 2.5m	Depth = 5.0m	Depth = 10.0m	Inside the hole
qu_{28} (predicted)	11.1	3.036	4.43	4.24	4.64	3.04
qu_{28} (measured)	29.52	4.282	4.28	4.28	4.28	4.28
ρ_0 (measured)	0.122	0.333	0.273	0.273	0.261	0.333
Equation used	1.2.7.A	10.13.16.A	10.13.16.A	10.13.16.A	10.13.16.A	10.13.16.A

3.2 Laboratory mixing experiment using site soil

Soil sample also collected from the same construction site in Arahama, Kashiwazaki-shi, Niigata prefecture. Sample was taken using from the bored hole depth of 21.0m (sandy soil) and 51.0m (gravel ground) and cement milk with water cement ratio 60%. Similar mixture experiment was conducted and the condition is listed in table 5. Electrical resistivity value is shown in Figure 13 and mixture P and Q has different initial electrical resistivity value compare other mixture. The influence of variation in gravel size of site soil is believed as a factor and temperature also has significant effect.

Table 5. Mixture condition for site soil experiment

W/C ratio (%)	Soil classification	Mud water amount ratio	Cement milk : mud water		Replacement soil percentage (%)
60	Sand	1.6	1:1	J	100
60	Sand	1.6	1:1.5	K	150
60	Sand	1.6	1:2	L	200
60	Gravel	1.7	1:1	P	100
60	Gravel	1.7	1:1.5	Q	150
60	Gravel	1.7	1:2	R	200

Compressive strength was also calculated using prediction formulation (equation 25 to equation 31) and obtained the same results for every mixture sample. This is believed caused by large error of predicted value when assumed ρ_0 was larger than soil electrical resistivity value. Based on this result, it is necessary to collect more data on the electric resistivity value estimation formula and modify it to an estimation formula incorporated to actual soil conditions.

4. Electrical Resistivity Measurement Based of Strength Prediction Recommendation

Based on performed laboratory experiment in this research, prediction formulation for compressive strength of bored pile construction using electrical resistivity measurement procedure is recommended by following diagram shown by Figure 14.

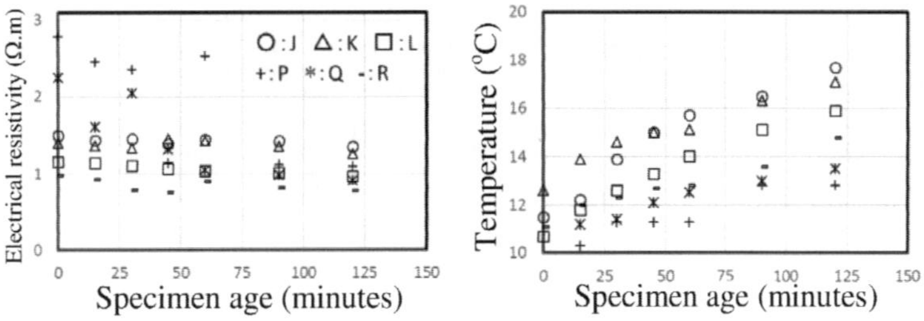

Figure 13. Electrical resistivity value and temperature change with time

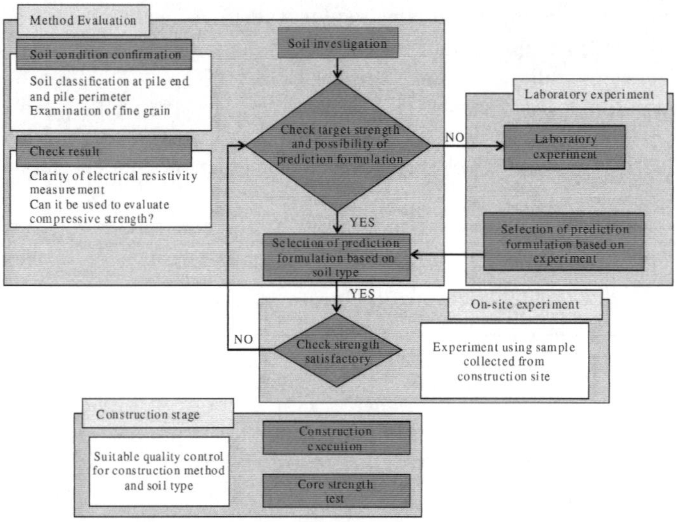

Figure 14. Recommendation procedure for compressive strength prediction soil cement mixture

5. Conclusion

1. Electrical resistivity measurement method 1 was the most suitable method to predict compressive strength of cement milk or cement soil mixture
2. Electrical resistivity value of specimen at early age tends to be consistent regardless to mixture configuration and cement amount if the particle size of soil is uniform to some extent
3. The temperature change of the specimen is greater than Atmospheric temperature has more influence to temperature change of the specimens rather than the heat emission due to the hydration reaction.
4. The relationship between qu_{28}/ρ_{28} and qu_3/ρ_3 is better than $\Delta qu/\Delta\rho$ and qu_3/ρ_3
5. It is possible to use measurement method 2 and 3 for laboratory experiment, however correction process is necessary.
6. Based on outdoor experiment, temperature change does not have significant effect to compressive strength.
7. It is necessary to perform another experiment to understand the influence of soil classification and particle size of soil to electrical resistivity value.
8. Equation 1 to equation 33 may be applied to predict compressive strength of actual condition, however accurate electrical resistivity measurement and measurement data accumulation is necessary.

6. References

[1] Tsuchiya T, Kuwabara F 2012 *Study on the Method of Constructing the Root-Hardened Part of the embedded Pile and its Strength Development* (Japan: Japan Architectural Institute Technical Report Vol. 18)
[2] Tsuciya T 2014 *Relationship Between the Strength of the Unsolidified Sample of the Root-Hardened Part and The Ground in the High Supportive Pile*, (Japan: The Architectural Institute of Japan Summary of Academic Papers Summary)
[3] Komatsu G, Fujii M 1998 *Quality Evaluation of Pile Tip Root-Hardened Part of Buried Pile Using Electric Resistivity*, (Japan: Summary of the Architectural Institute of Japan, page 621-622)
[4] Kato Y et al 2000 *Basic Study on Strength of Pile Fixed Portion of Embedded Pile*, (Japan: Summary of the Architectural Institute of Japan, page 755-756)
[5] Tomita K et al 2001 *Quality Evaluation of Solidified Body of Pile Fixing Solution*, (Japan: Summary of the Architectural Institute of Japan, page 693-694)
[6] Society of Civil Engineers: Concrete Standard Formulation Established in 1996 [Construction Edition]

IOP Conf. Series: Materials Science and Engineering **264** (2017) 012002 doi:10.1088/1757-899X/264/1/012002

Mix design and mechanical performance of geopolymer binder for sustainable construction and building material

Manfredi Saeli*, Rui M. Novais, Maria Paula Seabra and João A. Labrincha

Department of Materials and Ceramics Engineering, CICECO, University of Aveiro, Campus Universitário de Santiago, 3810-193 Aveiro, Portugal
*Corresponding author

Abstract. Sustainability in construction is a major concern worldwide, due to the huge volume of materials and energy consumed by this sector. Associated supplementing industries (e.g. Portland cement production) constitute a significant source of CO_2 emissions and global warming. Valorisation and reuse of industrial wastes and by-products make geopolymers a solid and sustainable via to be followed as a valid alternative to Portland cement. In this work the mix design of a green fly ash-based geopolymer is evaluated as an environmentally friendly construction material. In the pursuit of sustainability, wastes from a regional kraft pulp industry are exploited for the material processing. Furthermore, a simple, reproducible, and low-cost manufacture is used. The mix design is hence optimised in order to improve the desirable mechanical performance of the material intended for structural applications in construction. Tests indicate that geopolymers may efficiently substitute the ordinary Portland cement as a mortar/concrete binder. Furthermore, valorisation and reuse of wastes in geopolymers is a suboptimal way of gaining financial surplus for the involved industrial players, while contributes for the implementation of a desirable circular economy.

1. Introduction

Construction is a crucial sector for the economy and is in continuous growing worldwide, especially in the developing countries. Over the past century, ordinary Portland cement (OPC)-based concrete has become the highest volume manufactured product on Earth due to its extraordinary versatility, reliable performance and low cost processing technology. Unfortunately, many recent studies have demonstrated that its production is highly unsustainable, as the very large volumes required to meet global demand lead to a huge consumption of (non-renewable) raw materials and energy, while generate enormous emission of greenhouse gases, and dust pollution [1]. The global CO_2 emissions attributed to the production of OPC contribute to around 5-8% according to the most recent data [2-4]. In particular, OPC manufacture responds for 80% of the overall CO_2 release, while the production of aggregates contributes with 20%. The overall emissions of the concrete industry contribute as much as 10% of the global CO_2 releases [5, 6]. Such a large amount is highly unsustainable. As a consequence, the development of novel and more sustainable construction and building materials, with reduced environmental footprint, is currently a major concern in the worldwide construction industry.

Nowadays, geopolymers (GP) are recognized to be a viable and sustainable alternative to OPC. The first studies on GP, with direct applications in civil engineering, date back in the 1950s in the former Soviet Union where Portland cement shortage was affecting the country following the IIWW. Hence, Prof. G. V. Glukhovsky began to study the binders used in the ancient Egyptian and Roman structures and discovered the possibility of producing a novel kind of binders exploiting low basic calcium or calcium-free alumina-silicate materials mixed with solutions containing alkali metals [7, 8].

Subsequently, Prof. J. Davidovits developed and patented some binders obtained from the alkali-activation of metakaolin (MK) giving a strong boost to the scientific research [9].
GP are inorganic alkali-activated binders formed by the interaction of an alkaline solution with a reactive solid source of aluminium and silicon [10] that generates a three-dimensional structure of SiO_4 and AlO_4. Many solid precursors have been explored so far: ground granulated blast furnace slags, fly ashes (FA), although a range of non-blast furnace slags, natural pozzolanic materials, and calcined clays. The use of aluminosilicate-based industrial wastes might reduce the greenhouse footprint, in comparison to traditional concretes. Furthermore, under an accurate mix design it is possible to engineer novel GP-materials with comparable, or superior, mechanical and chemical properties than OPC [11]. Additionally, GP manufacture is highly cost effective and sustainable, since preparation and curing might occurs at room temperature and is simple and fast. Some benefits of GP-concrete are: short curing time, early strength gain, suitable pre-fabricated concrete products for fast-building construction, building securing, etc.. Moreover, the material shows excellent chemical resistance to harsh environment, to fire loads, and to abrasion, low shrinkage and thermal conductivity, etc. [12, 13]. All of that makes GP highly suitable either for construction materials and environmental uses. Many interesting applications were already reported in construction with potentials for future innovative developments. At present they include high strength concretes with good resistance to chloride penetration, fire and/or acid resistant coatings, and waste immobilization solutions for the chemical/nuclear industries. Other applications in construction, guaranteed by GP, are easy adhesion to fresh and old substrates of distinct nature (stone, steel, glass, etc.), high surface definition that might replicate mould patterns, and inherent protection of steel reinforcing. Finally, a great deal of research has been performing worldwide towards the structural behaviour of GP-concrete in load-bearing structural members such as reinforced concrete beams, columns, slabs, etc. [14, 15].
This paper investigates the mix designs for GP-based binders intended for structural purposes, and formulated with industrial wastes. Regional wastes generated from a kraft pulp industry are exploited as precursors for the material processing. The specific variabilities are outlined in relation to the specimens' mechanical performance. Despite the key technological attributes and the most innovative applications, environmental and industrial cost saving issues are the main drivers for the uptake of the investigated technology. Valorisation and reuse of industrial wastes and by-products is a compelling business for governments and industries in order to respect the strict parameters and regulations imposed by the international frameworks, as the Paris Agreement from the Climate Conference COP21 (2015) or the EU Climate Action [16]. Such an issue is central, not only for cement companies but also for other industrial sectors where a large amount of wastes and by-products generates yearly an elevated loss of financial resources for treatment and disposal, reducing *de facto* the overall profit margins. This paper discusses the practical, technical, environmental, and commercial drivers, as well as how GP technology may form part of an overall "Green Concrete" industry.

2. Materials

The mix design aims to produce a GP binder to be used as lime or cement replacement in structural applications in construction (final mortar classifiable at least as class M10, according to UNI EN 998-2). GP have been produced by a mixture of MK and biomass FA as a source of alumino-silicates. The MK is purchased as Argical™ M1200S from Univar®. Its chemical composition was evaluated by X-ray fluorescence (Panalytical Axios spectrometer) (table 1). FA have been extensively investigated and exploited in GP production [17, 18] as a partial MK replaced in the formulations.
The typology used in this work comes from a regional kraft pulp industry. FA chemical composition was obtained by X-ray fluorescence (table 1), while the particle size distribution was determined by laser diffraction (Coulter LS230 analyser) (table 2). The specific surface area (estimated by BET) was measured in 3.97 m^2/gr for FA, and 16.30 m^2/gr for MK.
The alkaline activation was granted by a mixture made of a solution (10M) of sodium hydroxide (Na Hyd.) (reagent grade, 97%, Honeywell) dissolved in distilled water and sodium silicate (Na Sil.) (Quimialmel LDA, D40 - PQ).

Table 1. X-ray fluorescence of fly ash and metakaolin (left).
Table 2. Particle size distribution of fly ash (right).

Oxides [wt%]	Fly ash	Metakaolin
SiO_2	38.47	57.71
CaO	16.73	0.11
Al_2O_3	14.79	41.03
K_2O	5.97	0.98
Fe_2O_3	5.94	1.77
MgO	3.44	0.29
SO_3	2.66	0.07
Na_2O	1.53	0.06
P_2O_5	1.12	0.06
TiO_2	0.76	1.84
MnO	0.50	0.01
LOI	6.39	1.88

μm	%<
10	3.73
25	13.32
50	34.17
75	61.30
90	105.31
mean	49.28

3. Geopolymeric binder manufacture and characterisation

Specimens were manufacture by a simple and reproducible mechanical process performed at environmental temperature (20°C) to avoid any supply of external heat and, consequently, reducing the costs. GP specimens were prepared as follows: a) homogenization of the Na Hyd. solution and the Na Sil. at 50 rpm for 5 mins; 2) mixture of the alkaline solution with the FA and MK at 100 rpm for 10 mins; 3) pouring the slurry into cubic (30x30x30 mm) plastic moulds sealed with a plastic film for 24 hours (20°C, 65% RH); 4) demoulding hardened samples and curing at room conditions up to 28 days. For each formulation, the compressive strength (CS) was determined using an automatic hydraulic press (Shimadzu, AG-A) according to UNI EN 1015-11:2007 and ASTM C109 up to 28 days. Data were reported as the average from four tests. For scanning electron microscopy analysis (SEM) a microscope (Hitachi S4100) equipped with energy-dispersive spectroscopy (EDS) was used. Water absorption was evaluated as a percentage of ΔW/Wi (W=weight) by specimens immersion in distilled water for 24 hrs once they had been previously dried to constant mass (UNI EN 1015-18:2004).

4. Mix design: optimization of fly ash amount and alkaline activators ratio

GP mix design was outlined to engineer the novel binder with a solid mechanical performance (structural application in construction, maximisation of breaking point in compression) along with a suboptimal incorporation of the industrial wastes. FA have been exploited in various quantities for partial replacement of MK; different alkaline activators' ratios have been investigated as well. The performed analyses and mechanical tests lead to a suboptimal GP-based binder composition.

Traditional binders (whether cementitious or lime-based) harden in contact with water generating hydrates (a process that may last years in the case of lime products). Furthermore, they show a large reduction in pore volume due to the consumption of water during the hydration reaction, with a subsequent great volumetric shrinkage [19]. On the contrary, GP-based binders do not form hydrates and the pore size distribution is refined throughout the hardening process [20, 21]. As a consequence, the ratio water/binder in GP should be as lower as possible to obtain a non-porous binder with a good mechanical performance. The use of a low liquid/solid ratio also reduces the tendency for efflorescence formation. Furthermore, the molar oxide ratios SiO_2/Al_2O_3, Na_2O/Al_2O_3, and Na_2O/SiO_2 should range within the Davidovits limits [9] returning also a good indication of such a possible on going disastrous phenomenon that raises concerns on the possible materials durability.

MK is the most used precursor in GP production due to its relatively low cost along with the capability to develop high strength products [22, 23]. FA have been found to be a valid alternative source of aluminium and silicon to MK and can be successfully employed.

Different mixes of MK/FA were designed in order to study the effect of the FA content on the GP final breaking point (table 3a). Afterwards, having set a suboptimal FA amount, to optimise the best alkali activators' ratio different ratios of Na Hyd. solution and Na Sil. were tested (table 3b). Subsequently, considering that larger FA particles might not be highly reactive, also due to their organic nature, a set of them was sieved at 75 μm and 125 μm (table 3c). These fractions were then used. That is aimed to maximise FA incorporation, along with reducing the samples' overall porosity.

Table 3. Mix design (in weight) of geopolymeric binders for structural applications.

Phase	Composition	MK	FA	Sodium hydroxide	Sodium silicate	note
(a)	I (75%)	1	3.0	2	2	
	II (72.97%)	1	2.7	2	2	
	III (70.59%)	1	2.4	2	2	
	IV (67.74%)	1	2.1	2	2	
	V (64.29%)	1	1.8	2	2	
	VI (60%)	1	1.5	2	2	
(b)	A	1	2.33	2	2	
	B	1	2.33	1.5	2.5	
	C	1	2.33	1	3	
	D	1	2.33	0.5	3.5	
(c)	C1	1	2.33	1.5	2.5	d<75 μm
	C2	1	2.33	1.5	2.5	75<d<125 μm
	C3	1	2.33	1.5	2.5	d<125 μm

5. Results and discussion: mechanical performance

For each formulation, the CS, along with the mechanical performance, was determined up to 28 days curing. Data were reported as the average from four tested cubes randomly taken from the batch.
The preliminary phase (a) was performed in order to define the best amount of FA to substitute MK, aiming to optimize CS values. Thus, the ratio FA/MK has been varied from 3:2 (60%) to 3:1 (75%). Figure 1 reports the CS values of specimens containing 75% and 70% FA. The CS variation is not linear during curing due to the progress of the reactions of geopolymerisation and densification changes of the gels. The final best performance was achieved for a 70% incorporation of FA, corresponding to 6.19±0.46 MPa. This formulation was then explored in the following steps. Water absorption and apparent bulk density have been calculated respectively as 31.12 % and 1405 kg/m^3 for the 75% FA sample, and 32.97 % and 1265 kg/m^3 for the 70% FA sample. Figure 2 shows the micrographs of the two specimens. 70% FA appears to be more compact and presents many needle-shaped formations of sodium salt that enhance the bonding character of the whole structure; hence a superior compressive strength was obtained. However, such a formations thereafter appear as efflorescence after immersion in water - as will be discussed - generating issues on durability.

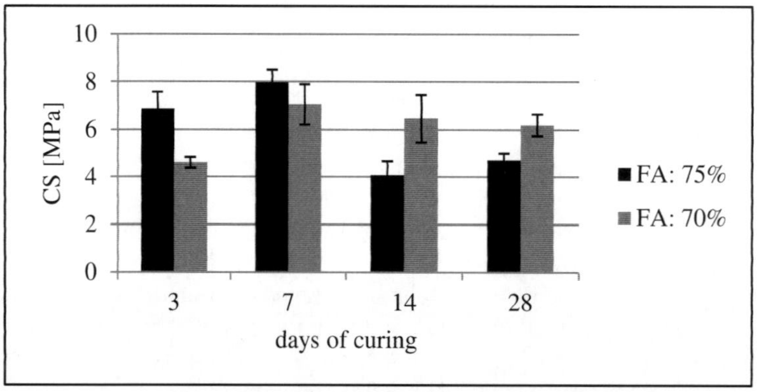

Figure 1. Compressive strength at 3, 7, 14 and 28 days curing for 75% e 70% fly ash content.

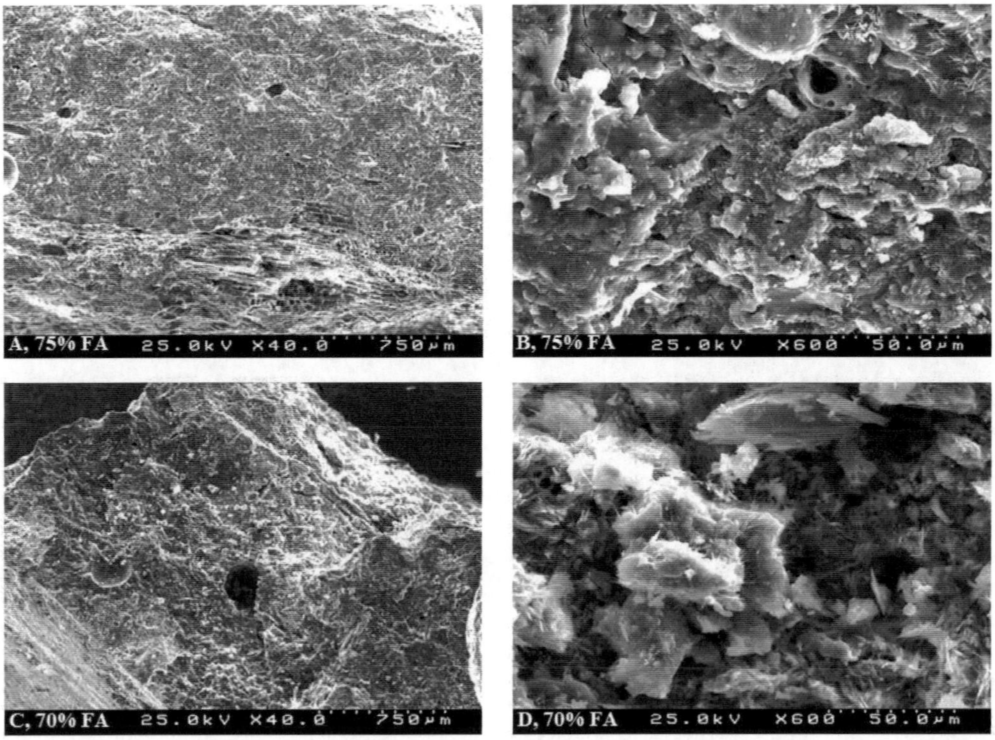

Figure 2. Micrographs at 750 μm and 50 μm, respectively A-B for 75 % FA (top images), and C-D for 70 % FA (bottom images).

Phase (b) is aimed to optimise the alkali activator formulation (Na Hyd. / Na Sil. ratio). Figure 3 shows the CS values of the specimens cured up to 28 days. As a result, the decrease of such a ratio enhances the mechanical strength. Values for 28-day cured samples are: A – 6.19±0.46 MPa, B – 7.86±1.23 MPa, C – 12.24±1.79 MPa, and D – 13.01±1.01 MPa. Values of the water absorption are: A – 31.12 %, B – 32.97 %, C - 38.73 %, and D – 26.59 %. Figure 4 shows the SEM pictures of mix C, revealing the presence of a larger number of small pores, in comparison to composition A (figure 2, C-D). Anyway, the overall gel looks denser, hence explaining the better mechanical performance. Table 4 reports the theoretical molar ratios of the four prepared mixes. Noteworthy that decreasing Na_2O to Al_2O_3 and SiO_2 ratios causes CS increase. Mix A contains the highest Na_2O concentration, responsible for the efflorescence appearance, as already stated. The usability of this material is then questionable. No efflorescence has been observed when Na Hyd. decreases. At the same time, a better slurry workability is achieved by increasing the H_2O content in the mixture. Although composition D shows the highest CS, at the same time it is highly cost ineffective as Na Sil. is more expensive than Na Hyd. It is also the least environmental interest. By balancing all observations, composition C represents the best compromise between mechanical performance and environmental/financial sustainability.

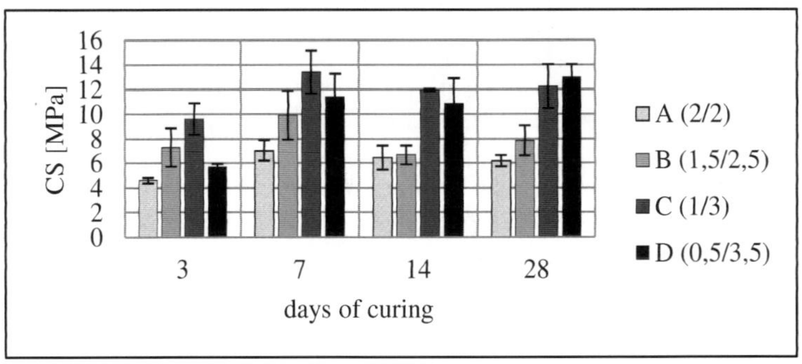

Figure 3. Compressive strength at 3, 7, 14 and 28 days curing for mix designs phase (b), A, B, C, and D. Ratios (Na Hyd./Na Sil.) are indicated.

Table 4. Theoretical molar ratios for phase (b) mix design.

	A 100/0	B 100/0	C 100/0	D 100/0	Davidovits limit
SiO_2/Al_2O_3	4.58	4.92	5.27	5.61	3.30 – 4.50
Na_2O/Al_2O_3	1.69	1.50	1.31	1.12	0.80 – 1.60
Na_2O/SiO_2	0.37	0.30	0.25	0.20	0.20 – 0.48
H_2O/Na_2O	12.9	14.2	15.9	18.2	10 – 25

Figure 4. Micrographs at 750 μm and 50 μm, respectively A and B, for the mix design C.

As already mentioned, phase (c) aims to study the effect of FA particle size. Finer particles tend to show higher reactivity, while improving the compactness of the samples. Both effect should increase the mechanical strength and reduce the penetration of aggressive external agents that affects the durability, as observed for ordinary artificial stone materials [24]. The use of finer materials demands extra water to correctly mix the components and adjust the rheology of the system. The water content should be carefully controlled, since an excess will create an extra volume of pores after curing. Figure 5 shows CS and the mechanical behaviours of mixes C_n using sieved FA. Tests clearly show that even if there is a slight improvement using sieved FA (d<75 μm), 12.77 MPa against 12.24 MPa, then the deviation is not that large to justify an extra employment of energy, and financial resources, to mill and sieve the material. Analogously, the overall mechanical performance, at least in the elastic region, is similar, meaning that FA particles size has a little influence in the overall behaviour.

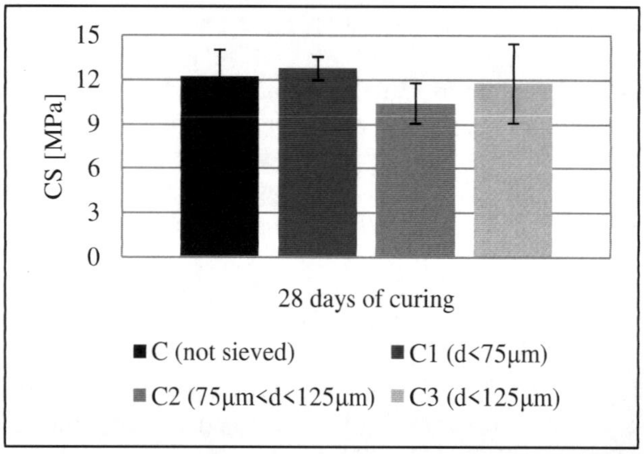

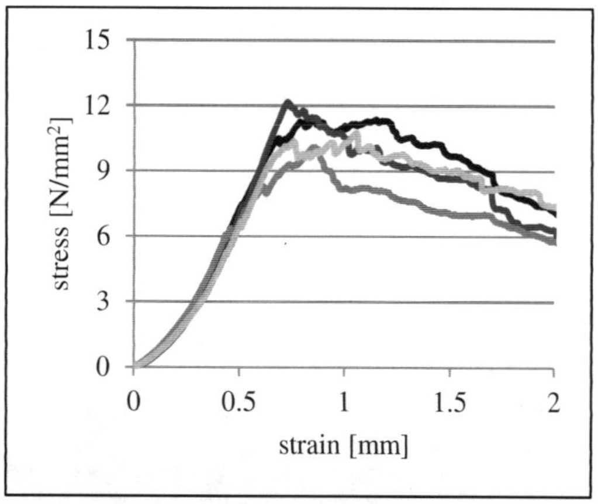

Figure 5. Compressive strength at 28 days curing for mix designs phase (c): left: break point; right: mechanical performance.

6. Standard mechanical tests

All the above discussed factors constitute the basis for a further test according to the standard UNI EN 1015-11:2007. Thus, geopolymer specimens were produced with formulation C in standard dimensions (40x40x160 mm). The standardized procedure prescribes that a mortar, in order to be classifiable as a structural material, might show a compressive strength at least equal to 10 MPa, class M10 according to UNI EN 998-2. The new performed mechanical tests have revealed a final breaking point equal to 22.15±1.22 MPa at 28 days curing. The final resistance to bending was also measured returning 3.05±0.37 MPa. Figure 6 shows the mechanical performance of the optimized geopolymeric binder at 28 days curing, following the prescribed standardized procedure. All considering, such a geopolymer, under the discussed formulation C, could be classifiable equal to a class M20 mortar.

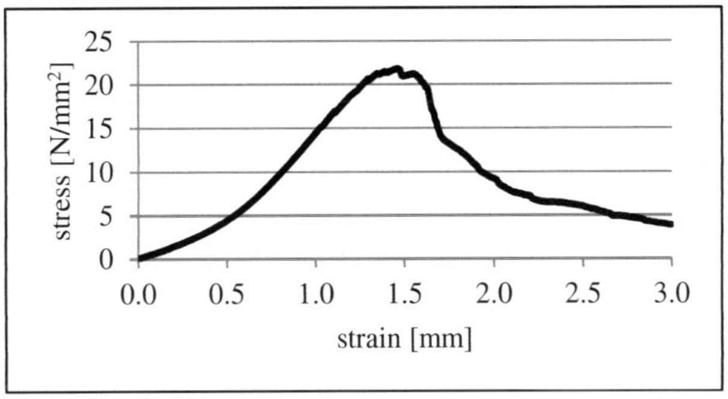

Figure 6. Mechanical performance of the optimized geopolymeric binder at 28 days curing, following the standard procedure.

7. Conclusion

This work analysed the possibility of producing novel geopolymeric binders by valorising and reusing some industrial wastes, such as the fly ash, from a regional kraft pulp industry. Different mix designs have been tested in order to optimise the final mechanical performance of the binder intended for structural applications in construction and a sustainable replacement of the ordinary Portland cement.
Based on the investigation carried out, it is concluded that, with these particular wastes, the incorporation of 70% fly ash in the geopolymer paste assures the best 28 days compressive strength. Furthermore, fly ash does not need any particular treatments, such as sieving, and can be incorporated as produced. Indeed sieving does not cause any relevant effects on the final compressive strength. The best alkaline activator presents a ratio sodium hydroxide/sodium silicate equal to 1:3 leading to a break point equivalent to 12.24±1.79 MPa, corresponding to the mix design C.
Moreover, the mechanical tests have been repeated following the standard for mortars, leading to a break point equivalent to 22.15±1.22 MPa and a bending resistance equal to 3.05±0.37 MPa. All these values are suitable for structural applications in construction, with a classification equal to M20.
All considering, the wastes from the kraft pulp industry can be opportunely reused to produce a green fly ash-based geopolymer intended for an environmentally friendly construction material. Furthermore, in the pursuit of sustainability, manufacture is performed at environmental temperature and is simple, reproducible, and low-cost leading to a surplus of profit margins for the industry.

8. Acknowledgement

This work is financed by Portugal 2020 through European Regional Development Fund (ERDF) in the frame of Operational Competitiveness and Internationalization Programme (POCI) in the scope of the project PROTEUS - POCI-01-0247-FEDER-017729 and in the scope of the project CICECO - Aveiro Institute of Materials CTM /50011 (Compete Refrence:POCI-01-0145-FEDER-007679), Associated Laboratory of University of Aveiro, co-financed by Fundação para a Ciência e Tecnologia/MCTES.

9. References

[1] Mehta PK 2010 *Sustainable cements and concrete for the climate change era – a review* in Zachar J, Claisse P, Naik TR and Ganjian E *Proceedings of the II Int Conference on Sustainable Construction Materials and Technologies* (Università Politecnica delle Marche) pp. 1–10
[2] Zhang Z, Provis JL, Reid A and Wanga H 2014 *Constr Build Mater* **56** 113–127

[3] Schneider M, Romer M, Tschudin M and Bolio H 2011 *Cem Concr Res* **41(7)** 642–50
[4] Flower D and Sanjayan J 2007 *Int J Life Cycle Assess* **12** 282–88
[5] Huntzinger DN and Eatmon TD 2009 *J Clean Prod* **17(7)** 668–75
[6] Meyer C 2009 *Cem Concr Compos* **31(8)** 601–05
[7] Palomo A, Krivenko P, Garcia-Lodeiro I, Kavalerova E, Maltseva O and Fernández-Jiménez A 2014 *Mater Construcc* **64** 315
[8] Provis JL and Van Deventer JSJ 2014 Alkali Activated Materials: State-of-the-Art Report (RILEM TC 224-AAM, NY Springer) p. 32
[9] Davidovits J 1979 *Synthesis of new high temperature geo-polymers for reinforced plastics/composites* (Society of Plastic Engineers, Brookfield Center) pp. 151–54
[10] Komnitsas K. and Zaharaki D 2007 *Miner Eng* **20** 1261-77
[11] Duxson P, Fernández-Jiménez A, Provis JL, Lukey GC, Palomo A and Van Deventer JSJ 2007 J Mater Sci **42(9)** 2917–33
[12] Zhang HY, Kodur V, Qi SL, Cao L and Wu B 2014 *Constr Build Mater* **55** 38-45
[13] Turner LK, Collins FG 2013 *Constr Build Mater* **43** 125–130
[14] Mo KH, Alengaram UJ and Jumaat MZ 2016 *Constr Build Mater* **120** 251–64
[15] Duxson P, Provis JL, Lukey GC and Van Deventer JSJ 2007 *Cement Concrete Res* **37** 1590–1597
[16] https://ec.europa.eu
[17] Al-Zboon K, Al-Harahsheh MS and Hani FB 2011 *J Hazard Mater* **188** 414-21
[18] Embong R, Kusbiantoro A, Shafiq N and Nuruddin MF 2016 *J Clean Prod* **112** 816-22
[19] Brouwers H 2004 *Cement Concrete Res* **34(9)** 1697–1716
[20] Duxson P, Provis JL, Lukey GC, Mallicoat SW, Kriven WM and Van Deventer JSJ 2005 *Colloid Surface* A **269(1–3)** 47–58
[21] Sindhunata L, Provis C, Lukey H and Van Deventer JSJ 2008 *Ind Eng Chem Res* **47(9)** 2991–99
[22] Davidovits J 1991 *J Therm Anal* **37(8)** 1633–56
[23] Duxson P, Mallicoat SW, Lukey GC, Kriven WM and Van Deventer JSJ 2007 *Colloid Surface* A **292(1)** 8–20
[24] Glasser FP, Marchand J and Samson E 2008 *Cement Concrete Res* **38(2)** 226–46

Effect of setting accelerator to the initial strength of mortar with blast furnace slag cement

D Takehisa[1], N Nakajima[2], Y Uno[3] and S Date[4]

1 Tokai University/Department of Civil Engineering, Hiratsuka City, Japan
2 Tokai University/Department of Civil Engineering, Hiratsuka City, Japan
3 Sato kogyo Co.,Ltd./Department of Civil Engineering, Atsugi City, Japan
4 Tokai University/Department of Civil Engineering, Hiratsuka City, Japan
E-mail: takehisa.tokai@gmail.com

Abstract. Effect of three types of setting accelerators such as calcium nitrous, calcium nitrate and C-S-H micro crystal dispersion on early strength enhancement of mortar with blast furnace slag cement were investigated. As a result, it was confirmed that compressive strength of mortar at early age had been enhanced by each accelerator, among them, the accelerator of C-S-H type showed the highest promoting effect. Furthermore, it was found that when the dosage is 4% or more, the initial strength designated by construction procedure can be secured and the productivity of construction work can also be improved.

1. Introduction

There are many mountain ranges and about 9000 tunnels in Japan. Various method for tunneling work had been developed so far. Among them, New Austrian Tunneling Method (NATM) and shield method are being used mainly. In tunnel construction site in Japan, the NATM is also being adopted for tunnel construction not only in mountains but also in urban areas [1]-[2]. In the construction process of the NATM, after excavating natural ground through after undergoing geological investigation such as boring, then shotcrete, a primary lining to fix the natural ground, is applied immediately. Afterwards, drill a hole with a punching machine, lock bolt, waterproof processing is carried out, and the secondary lining concrete is cast by using Centre for finishing. Center stands for the formwork of the arch part when placing the lining concrete, and it has an assembled and a movable mold frame. Besides, as an advantage of performing secondary lining, increasing the safety factor as a structure in consideration of uncertain factors such as geological inhomogeneities, quality of support workers, corrosion of lock bolts, etc. Moreover, the durability of structure can be improved against changes in external force after service start and deterioration of the ground and timbering members. Recently, utilization of blast furnace slag cement of which is excellent in long-term strength development, measures to reduce environmental burden and effective for suppressing alkali aggregate reaction, for the secondary lining concrete has attracted attention [3]-[7]. Blast furnace slag cement is one in which 30% to 70% of ordinary Portland cement is replaced with slag fine powder [8].
Blast furnace slag fine powder is quenched and finely pulverized molten blast furnace slag produced as a by-product in the production of pig iron and is an supplemental cementitious material having the property of being hydrated and hardened by alkali stimulation brought by calcium hydroxide which is provided through hydration of the cement, gypsum, etc. [9]-[11]. Concrete using blast furnace slag cement is slower in early strength development than ordinary cement-based concrete, and in order to exert performance such as designated strength, durability, crack resistance, etc., it is necessary to

conduct appropriately curing. It is generally known that when the initial wet curing is insufficient, it hinders the strength development [9]-[19]. Therefore, when blast furnace slag cement is used, there is a disadvantage that the construction term is likely to be extended. Among such conditions, the reason why blast furnace slag cement is being desired for the construction materials is that material cost is lower than ordinary cement and high durability is mentioned, and in order to prevent delays in the construction period when blast furnace slag cement is used in these construction work, an initial strength enhancement must be the key technology [20]-[24]. On the other hands, various concrete admixtures are currently on the market, but the current situation is that there is no achievement of use of these products in tunnel secondary lining.

In this study, an effect of three types of setting accelerators on the initial strength of the mortar with blast furnace slag cement under various curing temperature was evaluated and possibility of improvement of the productivity of construction work using blast furnace slag cement mixed concrete was also investigated.

2. Experiment summary

2.1. Materials used

The materials used are shown in TABLE Ⅰ. A polycarboxylic acid type high performance AE water reducing agent was used as a superplasticizer, and three types of setting accelerators, C-S-H micro crystal dispersion type and nitrous type A and B were used respectively. As the main component of the setting accelerators, the C-S-H type is a calcium silicate hydrate nanoparticle, the nitrous type A is an inorganic nitrogen compound, and the nitrous type B is a setting accelerators comprising both calcium nitrous and calcium nitrate.

Table 1. Materials used.

Material		Properties	Density (g/cm^3)
Cement	BB	Blast furnace slag cement Type B	3.04
Fine aggregate	S	River sand from Yamakita Kanagawa	2.69
Chemical admixture	Ad	High-performance AE reducing agent (Polycarboxylic acid base)	1.01~ 1.13
Setting accelerators	Ac	C-S-H micro crystal dispersion type	1.03~ 1.11
		Nitrous type A	1.41~ 1.45
		Nitrous type B	1.49

2.2. Experimental condition

Experimental parameter and curing conditions of the mortar are shown in TABLE Ⅱ. Materials used were kept at a prescribed temperature beforehand so that the mixing temperature was the same as the curing temperature, and the curing temperature was set to 10, 15 and 20 °C respectively. Incidentally, the curing temperature was kept constant, and no rising and falling gradients were provided in curing pattern. In addition, water cement ratio of the mortar was set to 42 %, which was optimal for material separation resistance and fluidity. In addition, the sand cement ratio was set to 3.0.

Experimental conditions are shown in TABLE Ⅲ. Total of 11 mix proportions were evaluated.

Table 2. Experimental parameter and curing conditions.

Parameter	Level
Setting accelerators	Additive-free
	C-S-H micro crystal dispersion type
	Nitrous type A
	Nitrous type B
Addition rate (C×%)	0, 2, 4, 6, 8
Curing temperature (℃)	10, 15, 20
Materials age (hour)	8, 10, 12, 14

Table 3. Experimental conditions.

NO.	W/C (%)	C-S-H (%)	Nitrous type A (%)	Nitrous type B (%)	Total addition amount (%)
1	42	0	0	0	0
2			2		2
3			4		4
4		2	0	2	
5			2	0	
6		4	0		
7				2	6
8			2	0	
9		6	0		
10				2	8
11			2	0	

2.3. Mortar mixing method

Figure 1. Hobart type mixer

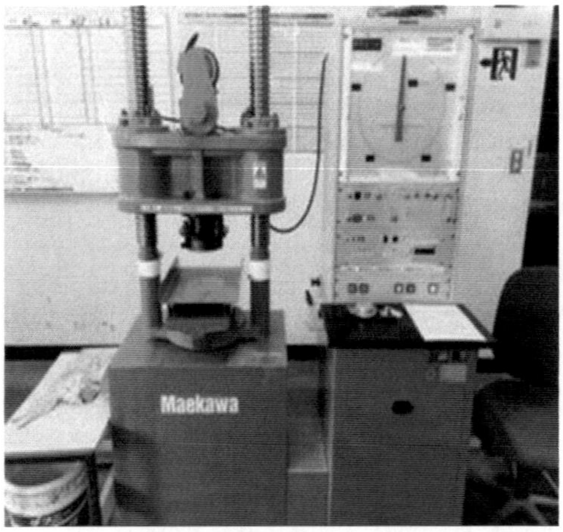

Figure 2. Compressive strength machine

A mixer used in this study, Hobart type mixer with a maximum capacity of 20 liters, is shown in the Figure 1 After adding cement and fine aggregate, mixing was carried out at low speed (110 r/min) for 30 seconds, immediately thereafter, admixture and water were added and mixed at low speed (110 r/min) for 60 seconds. Finally, mortar was mixed at high speed (230 r/min) for 30 seconds and was discharged.

2.4. Test items

The compressive strength test was conducted with the apparatus shown in the Figure 2. Material age is shown in TABLE III. Cylindrical specimens with φ50 × 100 mm were used for the test.

3. Results and discussion

The effect of the setting accelerators on the initial strength are shown in Figure 3 and Figure 4 for curing temperature at 10 °C and the results at 15 °C are shown in Figure 5 and Figure 6, and the results at 20 °C are shown in Figure 7 and Figure 8. From Figure 3 and Figure 4, it was confirmed that specimen without accelerator, as a control, can't be demolded by 14 hours, whereas in the case of dosing a setting accelerator, it was figured out that addition of the accelerator makes demolding of mortar possible in 8 hours. In addition, when 4% of setting accelerator was added regardless of type, the strength development of about 1 N/mm^2 was confirmed in 14 hours. When the addition rate was increased from 4 to 6%, the strength development of much more than 1 N/mm^2 was confirmed as well. From Figure 5 and Figure 6, in the case of control, when the curing temperature was 10℃, specified demolding strength was obtained with age of 14 hours. On the other hand, in case of curing temperature of 15 ℃, specified demolding strength was obtained at 8 hours. When various 4% of setting accelerator was added, strength development of about 2 N/mm^2 was confirmed at 14 hours. From Figure 7 and Figure 8, when 2% of nitrous type A was added, the compressive strength at 14 hours was 3.03 N/mm^2. Furthermore, when 4% of setting accelerators were added, a result of 4.0 N/mm^2 or more was confirmed in 14 hours. As the results, it was confirmed that performance of strength enhancement by C-S-H type exceeded to the other accelerators at the same dosage. There was no difference in the influence on promotion of the initial strength of the two kinds of nitrous type. Figure 9 shows the C-S-H type 4% addition rate. After curing at 10 °C, the compressive strength at 12 hours showed 0.69 N/mm^2. After curing at 15 °C, the compressive strength at 12 hours was 1.31 N/mm^2. After curing at 20 °C. The compressive strength at 12 hours was 3.78 N/mm^2. This indicates that the curing temperatures of 15 °C and 20 °C were markedly higher than the curing temperatures of 10 °C and 15 °C. Therefore, according to previous research, it can be considered that the initial strength easily increases as the curing temperature becomes higher [25]-[26].

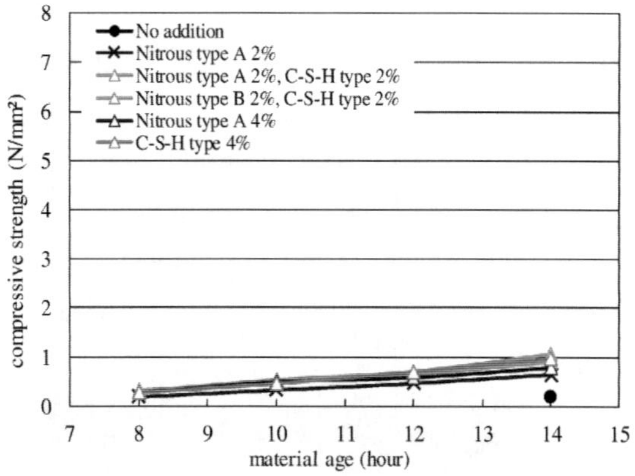

Figure 3. Material age vs compressive strength (curing temperature 10℃, Ac=4% or less).

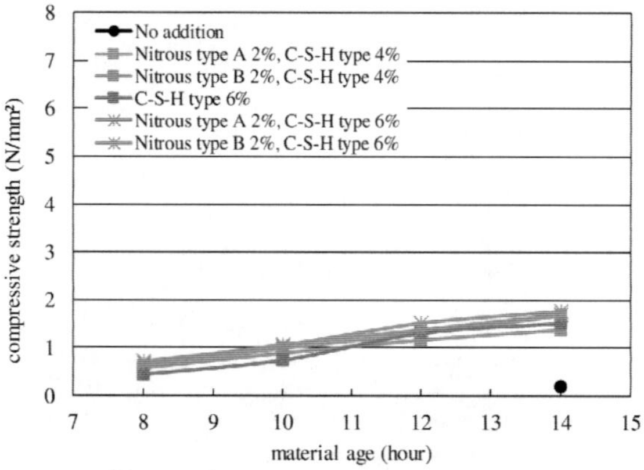

Figure 4. Material age vs compressive strength (curing temperature 10℃, Ac=6% or more).

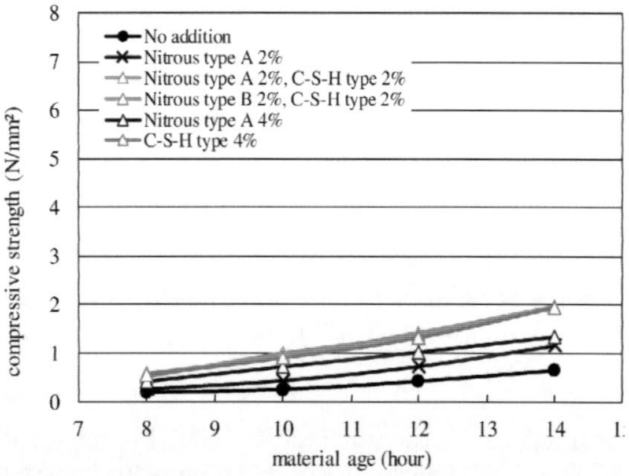

Figure 5. Material age vs compressive strength (curing temperature 15℃, Ac=4% or less).

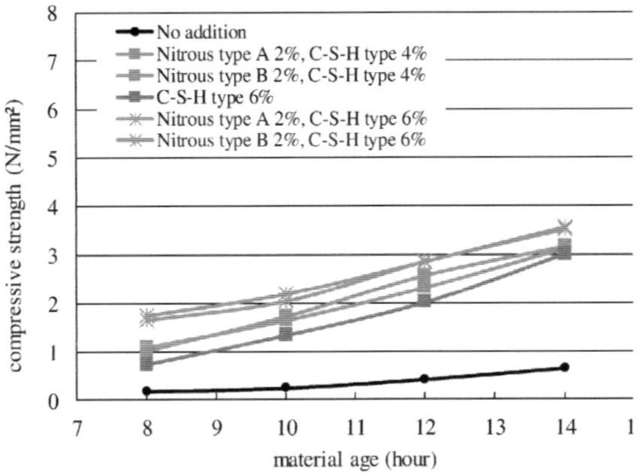

Figure 6. Material age vs compressive strength (curing temperature 15℃, Ac=6% or more).

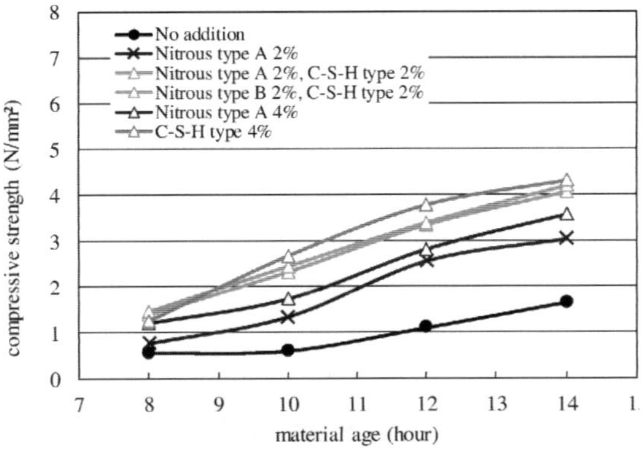

Figure 7. Material age vs compressive strength (curing temperature 20℃, Ac=4% or less).

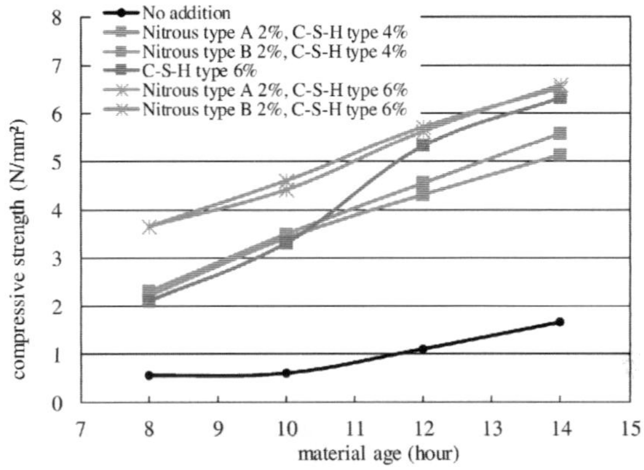

Figure 8. Material age vs compressive strength (curing temperature 20℃, Ac=6% or more).

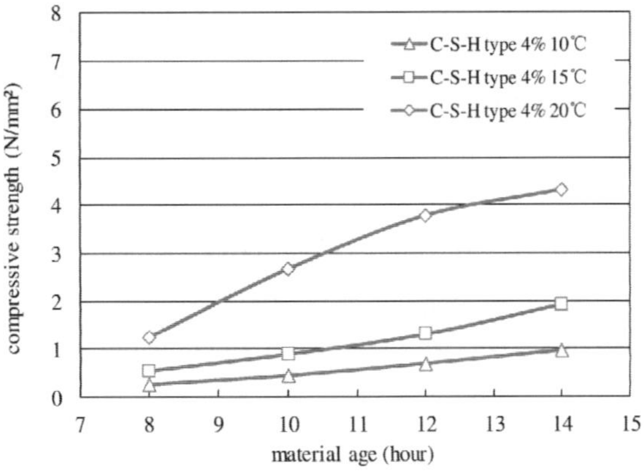

Figure 9. Material age vs compressive strength (Various curing temperature, Ac=4%).

It is generally known that the temperature inside the tunnel is around 20 °C [27]. In this study, when the curing temperature is at 20 °C, promotion of strength could be confirmed by adding a setting accelerators in compared to the plain sample. In addition, as a condition required at the work site to shorten the construction period, which is the improvement of productivity of blast furnace slag cement mixed concrete for tunnel lining, which is also the object of this research, the initial strength of 3 N/mm^2 or more within the age of 12 hours is necessary. With no addition, only the initial strength of 1.09 N/mm^2 had been confirmed in 12 hours of material age at 20 °C curing. When the addition rate of the setting accelerators was 4%, it was confirmed that the specified initial strength of 3 N/mm^2 or more was satisfied in 12 hours of the material age.

4. Conclusion

In this study, an effect of three types of setting accelerators on the initial strength of the mortar with blast furnace slag cement under various curing temperature was evaluated and possibility of improvement of the productivity of construction work using blast furnace slag cement mixed concrete was also investigated. As a result, the following was found.

1) When the curing temperature was 10 °C, the mold release strength was obtained in 14 hours. When the curing temperature was 15 °C, specified demolding strength was obtained in 8 hours.
2) Since the difference in compressive strength at curing temperatures of 15 °C,20 °C is significantly greater than the difference in compressive strength at curing temperatures of 10 °C,20 °C, the higher the curing temperature, the higher the initial strength.
3) Comparing the three kinds of setting accelerator, it was confirmed that the earliest intensity promotion was observed in the C-S-H type, and in the two types of nitrous type there was no difference in the influence on the promotion of the initial strength.
4) Materials used in the field require an initial strength of 3 N/mm^2 or more within 12 hours. Therefore, if the addition rate of the setting accelerator is 4% or more, the initial strength desired is satisfied.
5) When total addition rates are 4% or more regardless the type of accelerator, specified initial strength can be secured. Therefore, it leads to improvement of productivity and leads to shortening of construction period.

5. References

[1] Masayasu Hisatake: Basic study on stability, collapse behavior of the tunnel face, Japan Society of Civil Engineering memoirs, Vol.517, No.31, pp105-115, 1995-6.

[2] Frobel, R K: GEOSYNTHETICS IN THE NATM TUNNEL DESIGN, American Society of Civil Engineers Symposium on Geosynthetics for Soil Improvement at the ASCE Convention, pp51-67, 1988

[3] Hiroaki Koga, Tatsuo Niimi, Akinori Nakamura, Hiroyoshi Kato: Influence of the calcium hydroxide fine powder which gives it to the cement which contained blast furnace slag, cement concrete memoirs, Vol.67, No.1, pp151-156, 2013

[4] Katsuro Kokubu, Yoshiki Murata, Shigeru Takahashi, Hiroyuki Anzai: Study on insulation temperature rise and hydration property of the concrete using the blast furnace slag fine powder, Japan Society of Civil Engineering memoirs, Vol. 1988, No.369, pp39-48, 1988

[5] Takahiro Sagawa, Tetsuya Ishida, YaoLUAN, Toyoharu Nawa: Hydrate composition analysis and cavity structure properties of the Portland cement, Japan Society of Civil Engineering memoirs E, Vol.66, No.3, pp311-324, 2010-9.

[6] Ha-Won Songa, Velu Saraswathya: Studies on the corrosion resistance of reinforced steel in concrete with ground granulated blast-furnace slag, Journal of Hazardous Materials, Vol.138, No.2, pp226–233, 2006.11

[7] Hogan, F: The Effect of Blast Furnace Slag Cement on Alkali Aggregate Reactivity, Cement, Concrete and Aggregates, Vol.7, No.2, pp100-107, 1985

[8] Kazusuke Kobayashi, Akihiko Yoda: Use - as concrete-binding materials of Portland cement and the characteristic - blast furnace slag, concrete engineering, Vol.14, No.9, pp23-30, 1976

[9] Osborne GJ: Durability of blast-furnace slag cement concretes, Building Research Establishment Information Paper, IP6192, Building Research Establishment, 1992.

[10] R Doug Hooton: Canadian use of ground granulated blast-furnace slag as a supplementary cementing material for enhanced performance of concrete, Canadian Journal of Civil Engineering, 2000

[11] E. Douglas, J. Brandstetr: A preliminary study on the alkali activation of ground granulated blast-furnace slag, Cement and Concrete Research Vol.20, No.5, pp746-756, 1990

[12] J. Geiseler, H. Kollo, and E. Lang: Influence of Blast Furnace Cements on Durability of Concrete Structures, Materials Journal, Vol.92, No.3, pp252-257, 1995.5

[13] Kazuto Fukudome, Yukinori Furukawa, Akira Shono: Study on impact statement technique of the wet good self-care condition to give it to strength development of the concrete, Japan Society of Civil Engineering memoirs E2, Vol.67, No.1, pp18-27, 2011
[14] G.J. Osborne: Carbonation and permeability of blast-furnace slag cement concretes from field structures Proceedings, Third International Conference on Fly Ash, Silica Fume, Slag and Natural Pozzolans on Concrete, ACI Detroit, SP-114, Trondheim, Norway, Vol.2, pp.1209-1237, 1989
[15] M.D.A. Thomas, G.J. Osborne, J.D. Matthews and J.B. Cripwell: A comparison of the properties of OPC, PFA and GGBS concretes in reinforced concrete tank walls of slender section Magazine of Concrete Research 42, pp.127-134, 1990
[16] G.J. Osborne: The effect of external microclimate or the carbonation, permeability and strength of two blast-furnace slag cement reinforced concrete structures, Proceedings of P.K. Mehta Symposium on Durability of Concrete, Nice, France, pp.119-145, 1994
[17] K. Hollinshead, G.J. Osborne, D.J. Bigland: In service performance of reinforced concrete in the Tees Barrage. Durability of building materials and components, Proceedings of 7th International Conference, E and FN Spon, London, 1996
[18] K Hollinshead, G.J. Osborne, D.J. Bigland: durability assessment of concrete Proceedings of the Instr. Civil Engineers. Engineers, 115, pp.203-208, 1996
[19] G.J. Osborne: Carbonation of blast furnace slag cement concretes Durability of Building Materials, pp.81-96, 1986
[20] Y. Deng, X. Wu, M. Tang: High strength alkali-slag cement J. Nanjing Inst. Chem. Technol., pp.1–7, 1989
[21] C. Shi, X. Wu, M. Tang: Hydration of alkali-slag cements at 150 °C Cem. Concr. Res., pp.91–100, 1991
[22] J. Deja, J. Malolepszy: Resistance of alkali-activated slag mortars to chloride solution Proceedings of the Third International Conference on the Use of Fly Ash, Silica Fume, Slag and Natural Pozzolans in Concrete, SP-114, American Concrete Institute, Norway, Chicago, pp.1547–1561, 1989
[23] C. Shi, X. Shen, X. Wu, M. Tang: Immobilization of radioactive wastes with Portland and alkali-slag cement pastes Cemento, pp.97–108, 1994
[24] S. K. Srivastava V. K. Gupta Dinesh Mohan: Removal of Lead and Chromium by Activated Slag, Journal of Environmental Engineering, Vol.123, No.5, 1997
[25] Hisashi Sugiyama, Yoshihiro Masuda: Long-term strength development characteristics of the concrete which caught the initial high temperature history, Architectural Institute of Japan structure system memoirs, No.515, pp23-30, 1999
[26] Jotaro Morimoto, Taketo Uomoto: Study on hydration of the initial high temperature Porto land cement which I protected, concrete engineering annual collection of article reports, Vol.17, No.1, pp651-654, 1995
[27] Koji Baba, Tetsuo Ito, Hiromichi Shiroma, Kazuya Miyano, Hiroshi Nakajima, Yasushi Taniguchi: Tunnel pit environment and study on humidity change of the cover mechanic concrete, Japan Society of Civil engineering memoirs, Vol. which are undertaking construction 2003, No.742, pp27-35, 2003

IOP Conf. Series: Materials Science and Engineering **264** (2017) 012004 doi:10.1088/1757-899X/264/1/012004

Properties of ambient cured blended alkali activated cement concrete

M Talha Junaid[1]

1 Assistant Professor, Department of Civil and Environmental Engineering, College of Engineering, University of Sharjah, Sharjah, UAE
Email: mjunaid@sharjah.ac.ae

Abstract. This paper presents results of the development and strength properties of ambient-cured alkali activated geopolymer concrete (GPC). The study looks at the strength properties, such as compressive strength, splitting tensile strength, and elastic modulus of such concretes and its dependency on various parameters. The parameters studied in this work are the type and proportions of pre-cursor materials, type of activator and their respective ratios and the curing time. Two types of pre-cursor material; low calcium fly ash (FA) and ground granulated blast furnace slag (GGBFS) were activated using different proportions of sodium silicate and sodium hydroxide solutions. The results indicate that ambient cured geopolymer concrete can be manufactured to match strength properties of ordinary Portland cement concrete (OPC). The strength properties of GPC are dependent on the type and ratio of activator and the proportion of GGBFS used. Increasing the percentage of GGBFS increased the compressive and tensile strengths, while reducing the setting time of the mix. The effect of GGBFS on strength was more pronounced in mixes that contained sodium silicate as activator solution. Unlike OPC, ambient-cured GPC containing sodium silicate gain most of their strength in the first 7 days and there is no change in strength thereafter. However, GPC mixes not containing sodium silicate only achieve a fraction of their strength at 7 days and extended curing is required for such concretes to gain full strength. The results also indicate that the elastic modulus values of GPC mixes without sodium silicate are comparable to OPC while mixes with sodium silicate have elastic modulus values much lower than ordinary concrete.

1. Introduction

With a recent shift of focus in the construction industry to move towards a low carbon future, efforts are being made to find viable and effective alternatives to OPC concrete. Currently, concrete is the most used commodity second only to water with an average global consumption of 1m3 per person [1] and is responsible for the production of 5% of the global CO2 emissions in developed countries while as high as 10% in developing countries [1, 2]. To put these numbers in perspective, the total greenhouse gas emission by a highly industrialized country like Germany is documented in the Paris Agreement Article 21 Annex I to be 2.56% of the global emissions [3].

In the last decade, or so, extensive research has been undertaken to find alternates to Ordinary Portland Cement (OPC) concrete for construction. Among other alternatives, Alkali Activated Materials (AAM) [sometimes referred to as geopolymer concrete] have come to the forefront and offer a viable alternate to OPC. AAMs can be formed by reacting a silica (SiO2) and alumina (Al2O3) rich material, such as metakaolin, Fly Ash (FA), Silica Fume (SF), Rice Husk Ash (RHA) or Slag with alkali liquids such as a soluble metal hydro-oxide and/or alkali silicate [4-6]. Such binders combined with traditional coarse and fine aggregates to form concrete with fresh and hardened properties similar to or even superior to OPC. Consequently, and contrary to OPC, AAM concrete does not depend on calcining calcium

carbonate, which is the major source of CO2 emission in OPC concrete [7], and hence can reduce emissions by 45 to 80% for each cubic metre of OPC concrete replaced [6, 8].
Despite the advantages, there are several challenges that need to be overcome for AAMs to be used as a mainstream construction material. One such challenge is the dependency of the properties of these binders on the pre-cursor source and the activation system. Broadly AAM are categorized as low and high calcium system with hybrid systems gaining popularity in recent years [9].
Several studies on the mechanical behaviour of heat cured AAM are available in the literature [9-19], however, work on ambient cured AAM is rather limited. For heat cured AAMs the mechanical properties and microstructure of the matrix is highly influenced by the mixing conditions, rest period, curing time and temperature, and the concentration of the activating solutions. Nath and Sarker [19] recently reported on the modulus of elasticity and the flexural strength of ambient cured blended geopolymer concretes. They also proposed analytical relationships between experimental and predicted values of modulus for such materials. They concluded that the modulus of elasticity of ambient cured blended AAMs is 25-30% lower than that of OPC with comparable strength. The effect of moisture condition on the strength gain of heat and ambient cured geopolymer concrete was studied by Jiting and Kayali [20], while Nath and Sarker studied the effect of OPC on the early strength properties of geopolymer concrete cured at room temperatures [21]. Despite these relatively recent studies, there is an obvious need for further investigation on the various parameters that effect the properties of ambient cured AAMs. The work reported here is part of a wider investigation into the study of development and characterization of ambient cured blended AAMs. Initial findings of this investigation include the strength development and its dependency on the proportion of the precursor blend and the activator used.

2. Experimental work

2.1. Materials

Locally acquired low calcium Class F fly ash and GGBFS was used as pre-cursor material in this study. The composition of the pre-cursors materials (FA and GGBFS) used in this work, as found using X-Ray Fluorescence (XRF) are tabulated in Table 1. Sodium silicate and sodium hydroxide solutions were used as alkaline activators. Sodium silicate solution (Na_2SiO_3) with density between 1.296 and 1.396 gram/ml produced by Merck, KGaA, Germany and procured from ESTS, UAE was used. Sodium Hydroxide solutions (12M) were made from 98% purity flakes supplied by ESTS, UAE. Locally sourced, natural crushed aggregates, nominal sizes 20 and 10mm was used as coarse aggregates (with specific gravity of 2.68 and absorption of 0.7%), while a mixture 60%-40% of 5mm crushed aggregates and dune sand was used as fines (with specific gravity of 2.62 and absorption of 1%) in the mix.

Table 1. X-ray fluorescence (XRF) of fly ash and GGBFS used.

Material	Na_2O	MgO	Al_2O_3	SiO_2	SO_3	K_2O	CaO	TiO_2	MnO_2	Fe_2O_3	NiO	ZnO	SrO
FA	1.045	7.51	10.97	59.83	1.545	1.42	8.75	1.08	0.175	7.525	0.03	0.04	0.08
GGBFS	-	6.58	5.71	32.36	2.38	-	51.01	0.64	0.73	0.59	-	-	-

2.2. Specimen preparation and mix proportions

Fine and coarse aggregates in SSD condition were mixed with fly ash and GGBFS for 2 minutes. Silicates solution, if any, was then added to the dry mix and mixed for 3 minutes. Hydroxide was then added and mixed for 2 more minutes. Additional water along was then added and mixed for additional 2 minutes. The mix proportions used in this study are summarized in Table 2 and are based on the work of Junaid et al [22]. For all tested mixes, the water to binder ratio (w/b)was kept constant at 0.32. The ratio was calculated using the following equation:

$$w/b = \frac{W_{free} + W_{OH} + W_{Si}}{FA + G + S_{OH} + S_{Si}} \quad (1)$$

Where;

Wfree = Weight of free water in the system
WOH = Weight of water content of the OH solution
WSi = Weight of water content of the silicate solution
G = Weight of GGBFS
SOH = Weight of solid content of the OH solution
SSi = Weight of solid content of the silicate solution

Cylinders with nominal measurements of 100mm dia and 200mm height were mixed and casted. The specimen were wrapped in watertight membrane to minimize water loss during setting. The moulds were removed after 24 hours and the samples were placed in an environmental room with RH of 40% and temperature of 25°C till the time of testing. Compressive strength, splitting strength, and modulus of elasticity tests were performed using the ASTM C39/C39M-14 [23], ASTM C 496/C 496M-04 [24], and ASTM C469/C469M-10 [25] standards, respectively.

Table 2. Mix proportions of the AAM used in the study.

Mix ID		SET 1					SET 2				
		S1M1	**S1M2**	**S1M3**	**S1M4**	**S1M5**	**S2M1**	**S2M2**	**S2M3**	**S2M4**	**S2M5**
		kg/m^3	kg/m^3	kg/m^3	kg/m^3	kg/m^3	kg/m^3	kg/m^3	kg/m^3	kg/m^3	kg/m^3
Coarse	20 mm	475	475	475	475	475	475	475	475	475	475
	10 mm	590	590	590	590	590	590	590	590	590	590
fine	Dune	260	260	260	260	260	260	260	260	260	260
	5mm	390	390	390	390	390	390	390	390	390	390
NaOH	(12M)	164	164	164	164	164	82	82	82	82	82
Na_2SiO_3		0	0	0	0	0	82	82	82	82	82
FA		425	340	212.5	85	0	425	340	212.5	85	0
GGBFS		0	85	212.5	340	425	0	85	212.5	340	425
Water		58	58	58	58	58	66.5	66.5	66.5	66.5	66.5

3. Results and discussion

3.1. Setting time

The setting time of the mixes was determined using the Vicat needle apparatus and the final setting time is reported in Table 3. As is evident from the table that the setting times of mixes in Set 1 (without the silicate solutions) was generally higher than the corresponding mixes in Set 2 which had silicates and hydroxide in equal measure. Within a set, the mixes with higher amounts of GGBFS resulted in quicker setting times. The mix S1M1 was under observation for over 360 minutes, however a final setting time was not determined. This was due to the fact that this mix did not harden and the Vicat needle penetration readings did not change significantly during this time.

Table 3. Setting time of the tested mixes.

Sample ID	SET 1					SET 2				
	S1M1	S1M2	S1M3	S1M4	S1M5	S2M1	S2M2	S2M3	S2M4	S2M5
Final Setting Time (min)	360+	70	20	15	NT	120	65	15	10	NT

NT = not tested

Based on the above results, it can be concluded that an increase in the percentage of GGBFS, even in small amounts, decreases the setting time of the AAM mixes, significantly. This effect is more pronounced in mixes with equal amounts of silicates and hydroxides. The AAM mixes can therefore be designed to suit various application requirements as far as setting time is concerned. The setting

IOP Conf. Series: Materials Science and Engineering **264** (2017) 012004 doi:10.1088/1757-899X/264/1/012004

time of AAMs can be altered by varying the amounts and proportions of silicates, hydroxides, FA and GGBFS used in the mix for that specific purpose.

3.2. Strength

3.2.1. Compressive strength. The average 7-day compressive strength of the cylinders are given in Figure 1. For ease of readability in the results section the samples are identified using the following nomenclature:

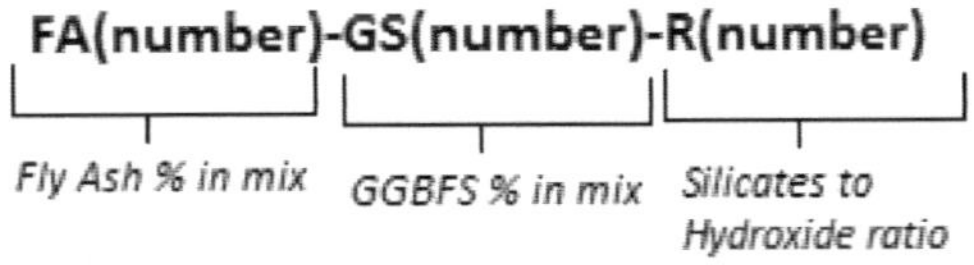

For Set 1 (with only hydroxide activator; R=0) the compressive strength increased as the percentage of GGBFS increased reaching a maximum value of just 16 MPa for 100% GGBFS at 7 days. However, the rate of change of strength with increasing GGBFS percentage is rather slow, with the only noticeable difference between the samples S1M1 and S1M2 (with 20% replacement of FA with GGBFS). The strength of the sample with no GGBFS (S1M1) was markedly lower than all the other tested specimen indicating that for single-activator alkali activated systems the use of fly ash alone is not advisable. Moreover, for such mixes the strength development is not significantly affected by the percentage of GGBFS after 20%. For all samples in this set, the early strength is only a fraction required for practical structural applications.

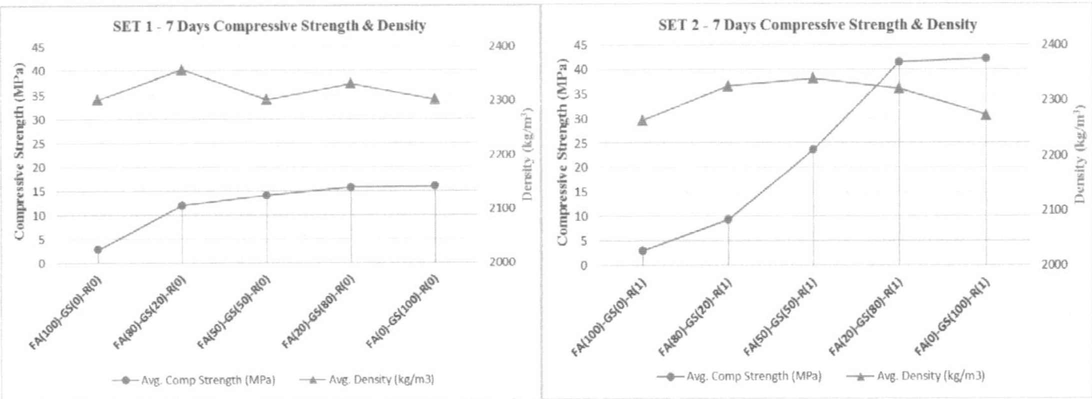

Figure 1. 7-day compressive strength of the samples for Set 1 and 2.

The average 7-day compressive strength of Set 2 (with silicate to hydroxide solution ratio R of 1) did not follow a similar trend as Set 1, with the most noticeable difference that the average strength values were significantly higher when compared to Set 1. The samples in this set returned a maximum strength value of just over 42 MPa at 7 days. The rate of strength development for this set was also much steeper than Set 1. When the GGBFS percentage was increased from 20% to 50% the strength also increased from 9.3 MPa to 23.6 MPa, an increase of just over 2.5 times. With a further increase of GGBFS from 50% to 80% the 7-day strength also increased from 23.6 to 41.5 MPa, an additional increase of over 1.7 times. Interestingly these increases in strength are in the same ratio as the increase in the GGBFS percentages of the mixes (from 20% to 50% and from 50% to 80%, an increase of 2.5 and 1.6 times, respectively). A further increase in the proportion of GGBFS had an insignificant effect on the 7-day strength of the specimen.
Nevertheless, the strength values of S2M1 and S2M2 were considerably lower than other samples in this set. It may, therefore, be concluded that for low proportions of GGBFS the early strength of all specimen tested is quite low and these mixes are not suitable for practical structural engineering applications at an age of 7 days.

The average 28 day compressive strength of samples in both sets are given in Figure 2 and they follow a similar trend to the strength acquired at 7 days. In general the samples in Set 1 returned a lower value of strengths with the maximum of under 25 MPa. Of interest is the strength gain of the samples in Set 1 at 28 days when compared to 7day strengths, with all samples gaining substantial strength, while those in Set 2 exhibit a slight gain only. Earlier work by researches [26, 27] suggest that AAMs gain all their strength in the first few days after casting and no further strength gain is expected after this initial period. However, the current finding may point to the dependency of the strength gain with time on the type of activator used and the precursor materials and its proportions, and further work needs to be done to reach a definite conclusion.
It is observed that the density of the alkali activated material is not significantly affected by the change in GGBFS percentage or the activator system. The average density of all samples was around 2300kg/m^3, which is comparable to ordinary concrete. The variation in density does not have a measurable effect on the strength of the specimen from both sets.

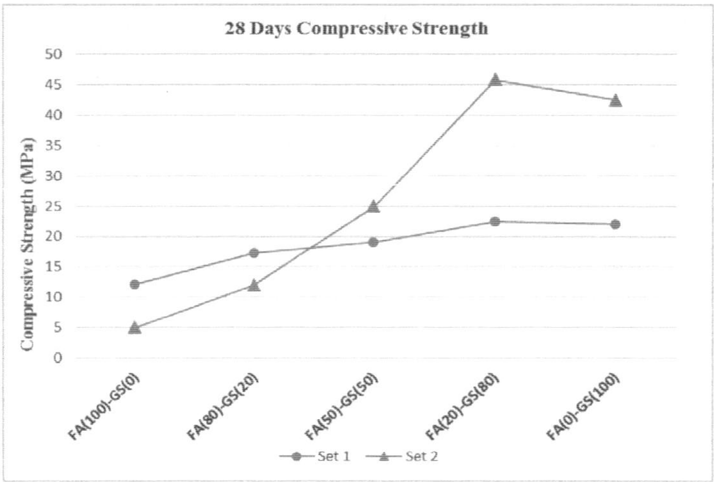

Figure 2. 28-day compressive strength of the samples for Set 1 and 2.

3.2.2. Splitting strength. The evolution of the splitting strength of AAM mixes of the two Sets of AAM samples are shown in Figure 3. As expected the splitting strength of the specimen follow the same trend as observed in the compressive strength development. The splitting strength of all samples at 7 days are approximately 10% of its compressive strength, which is widely used as a thumb-rule for determining the splitting strength of ordinary concrete. The maximum splitting strength at 7 days developed for Set 1 was just over 2.5 MPa, while that for Set 2 is 3.75 MPa. As is the case with compressive strengths, the tensile strength of Set 1 increased at 28 days when compared to 7 day values. However, the strength gain was not in the same proportion as for the compressive strength for this set. The tensile strength of Set 2 did not change with time, and follows the same trend as the compressive strength of this set.

3.3. Elastic modulus

Determination and prediction of elastic modulus of a building material is of immense importance as it is the primary factor in determining the flexural and axial stiffness of members. Comparative to OPC concrete, little work is reported on the elastic modulus of ambient cured AAM. There is a general agreement between all reported results that the elastic modulus of AAMs is lower than that of OPC with comparable strengths. The 28 day modulus of elasticity for the tested specimen (Set 1 and Set 2) are shown in Figure 4.

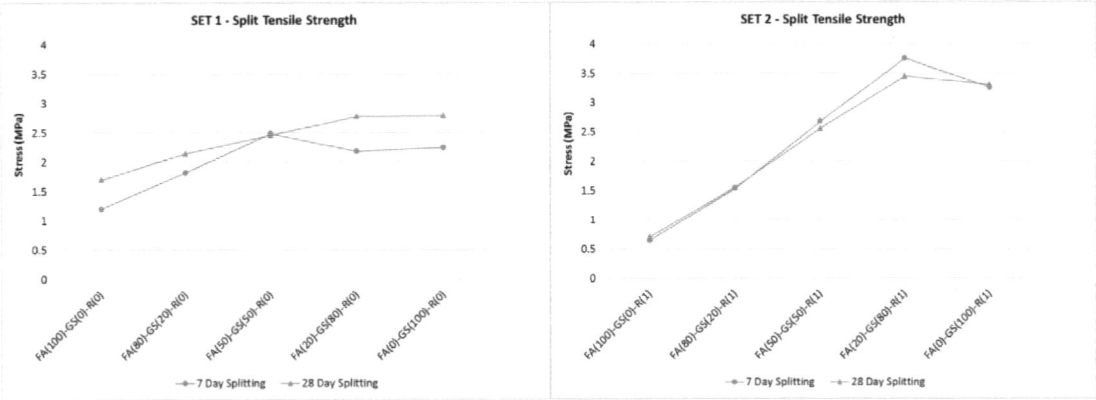

Figure 3. 7 and 28-day splitting strength of the samples for Set 1 and 2.

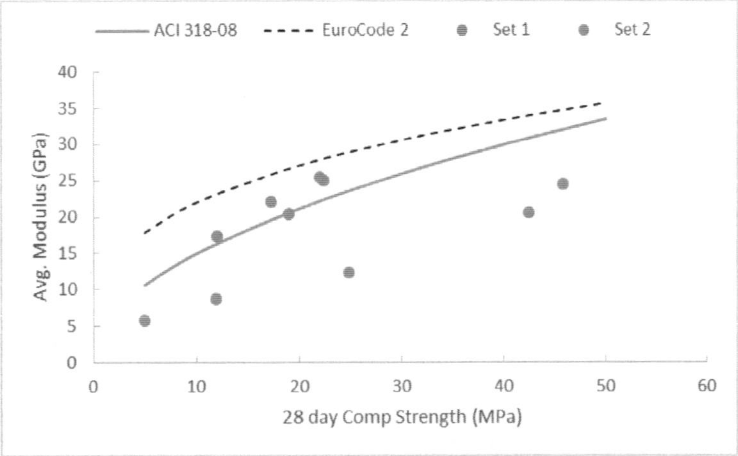

Figure 4. Young's modulus of AAMs as a function of compressive strength for Set 1 and 2.

Set 1 has a higher modulus value when compared to the samples of Set 2. These moduli values of Set 1 are a good fit with the prediction models for elastic modulus of OPC as put forward by ACI318-14 and EuroCode 2. However, the moduli values for the samples in Set 2 are much lower, although the samples reached a much higher strength. The only difference between the two sets is the presence of sodium silicate while all other parameters including the type and proportion of materials is the same. It may therefore be concluded that although sodium silicate has a positive effect on the strength of the AAMs, it may result in reduction of the elastic modulus. Also of note is the fact that both design code overestimate the modulus values of AAM containing silicates. It may therefore be concluded that standard equations can be used to predict the modulus values of AAM systems without silicate while other modified equations are required to accurately predict the modulus of AAM with silicates.

Junaid et al. [28] proposed a prediction equation for AAMs which is given in equation (2) below and demonstrated that this equation predicts with good accuracy the elastic modulus of AAMs.

$$E_c = 0.015 f_c'^{\,0.84} \rho^{1.43} \quad (2)$$

Where fc' is the 28 day compressive strength (in MPa) and r is the density of the alkali activated concrete. The results of these tests are plotted using equation (2) and are presented in Figure 5. The results clearly show that the prediction model presented by Junaid et al can accurately predict the Modulus values of AAMs containing silicates and hydroxide as activator (Set 2), while it under predicts the modulus of AAMs activated using only hydroxides (Set 1). As Set 1 returned strength values that are rather low and generally not suitable for structural applications, this proposed model presented above appears to be a suitable alternate for predicting AAM modulus values.

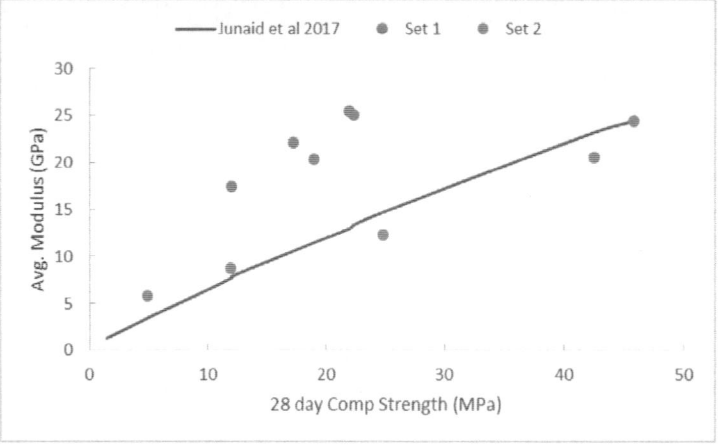

Figure 5. Young's modulus of AAMs as a function of compressive strength.

4. Summary and conclusions

A testing program was undertaken to determine the strength characteristics of ambient cured AAMs and their dependency on the various parameters such as precursor type and proportion, activator type and ratios, and curing time was studied. The compressive and splitting strengths along with modulus values were determined for different mixes. Based on the results of this study the following conclusion can be drawn:

- The strength properties of ambient cured alkaline activated binder concrete are greatly influenced by both the type and ratios of the precursor material and the activator solution. Addition of sodium silicate to the AAM system generally increases the strength of the binder but decreases the modulus of the resulting concrete.
- The splitting tensile strengths of AAM's is only a fraction of its compressive strength value and follows the same trends as the compressive strength. For systems with sodium silicate the splitting tensile strength is not effected by the curing time after the initial 7 days.
- The density of ambient cured AAMs is comparable to that of OPC concrete.
- Setting time of AAMs can be controlled by varying the proportion of GGBFS and sodium silicate in the system. Addition of both these compounds reduce the setting time of AAMs.
- The elastic modulus of AAMs is usually lower than OPC of comparable strength. It is therefore not recommended to use prediction models for OPC to predict the moduli of AAMs. The model proposed by Junaid et al [28] give accurate prediction value of the elastic modulus and is recommended to be used instead.

5. References

[1] Davidovits, J., *Global Wraming Impact on the Cement and Aggregate Industries.* World Resourse Review 1994. **6**(2): pp 263-278

[2] Worrell, E., et al., *Carbon Dioxide Emissions from the Global Cement Industry.* Annual Review of Energy and the Environment, 2001. **26**: pp 303-329

[3] Nations, U. *Paris Agreement.* 2016; Available from: http://unfccc.int/paris_agreement/items/9986.php

[4] Davidovits, J. *Soft Mineralogy and Geopolymers.* in *Proceedings of the Geopoolymer 88 International Conference.* 1988 France

[5] Davidovits, J., *Geopolymers Inorganic Polymerie New Materials.* Journal of Thermal Analysis, 1991. **37**: pp 1633-1656

[6] Lloyd, N. and B. Rangan, *Geopolymer Concrete with Fly Ash,* in *Second International Conference on Sustainable Construction Materials and Technologies* 2010 Italy

[7] McCaffrey, R., *Climate Change and the Cement Industry*, in *Global Cement and Lime Magazine* 2002. pp 15-19

[8] Fernández-Jiménez, A., et al., *New Cementitious Materials Based on Alkali-Activated Fly Ash: Performance at High Temperatures.* Journal of the American Ceramic Society, 2008. **91**(10): pp 3308–3314

[9] Provis, J. and J. van Deventer, *Alkali Activated Materials State-of-the-Art Report*, J. Provis and J.S.J. van Deventer, Editors. 2014, RILEM

[10] Diaz-Loya, E.I., E. Allouche, and S. Eklund, *Factors Affecting the Suitability of Fly Ash as Source Material for Geopolymers.* Fuel, 2010. **89**: pp 992-996

[11] Diaz-Loya, E.I., E. Allouche, and S. Vaidya, *Mechanical Properties of Fly-Ash-Based Geopolymer Concrete.* ACI Materials Journal, 2011. **108**(3): pp 300-306

[12] Duxson, P., et al., *Geopolymer Technology: the Current State of the Art.* Journal of Material Science, 2007. **42**: pp 2917–2933

[13] Fernández-Jiménez, A., A. Palomo, and C. López-Hombrados, *Engineering Properties of Alkali-Activated Fly Ash Concrete.* ACI Materials Journal, 2006. **103**(2): pp 106-112

[14] Junaid, M.T., O. Kayali, and A. Khennane, *Response of Alkali Activated Low Calcium Fly-Ash based Geopolymer Concrete Under Compressive load at Elevated Temperatures.* Materials and Structures, 2016. **In Press**

[15] Junaid, M.T., A. Khennane, and O. Kayali, *Performance of Fly Ash Based Geopolymer Concrete Made Using Non-Pelletized Fly Ash Aggregates After Exposure to High Temperatures.* Materials and Structures, 2015. **48**(10): pp 3357-3365

[16] Junaid, M.T., A. Khennane, and O. Kayali, *Stress-Strain Behaviour of Alkali Activated Fly Ash Concrete at Elevated Temperatures*, in *2nd International Conference on Advances in Cement and Concrete Technology in Africa*2016: Tanzania. pp 301-308

[17] Junaid, M.T., et al., *Aspects of the Deformational Behaviour of Alkali Activated Fly Ash Concrete at Elevated Temperatures.* Cement & Concrete Research, 2014. **60**: pp 24-29

[18] Lee, N. and H. Lee, *Setting and Mechanical Properties of Alkali-Activated Fly Ash/Slag Concrete Manufactured at Room Temperature.* Construction and Building Materials, 2013. **47**: pp 1201-1209

[19] Nath, P. and P.K. Sarker, *Flexural Strength and Elastic Modulus of Ambient-Cured Blended Low-Calcium Fly Ash Geopolymer Concrete.* Construction and Building Materials, 2017. **130**: pp 22-31

[20] Xie, J. and O. Kayali, *Effect of Initial Water Content and Curing Moisture Conditions on the Development of Fly Ash-Based Geopolymers in Heat and Ambient Temperature.* Construction and Building Materials, 2014. **67**: pp 20-28

[21] Nath, P. and P.K. Sarker, *Use of OPC to Improve Setting and Early Strength Properties of Low Calcium Fly Ash Geopolymer Concrete Cured at Room Temperature.* Cement and Concrete Composites, 2015. **55**: pp 205-214

[22] Junaid, M.T., et al., *A Mix Design Procedure for Low Calcium Alkali Activated Fly Ash-Based Concretes.* Construction and Building Materials, 2015. **79**: pp 301-310

[23] Standard, A., *Standard Test Method for Compressive Strength of Cylindrical Concrete Specimens*, 1997

[24] Standard, A., *Standard Test Method for Splitting Tensile Strength of Cylindrical Concrete Specimens*, 1984, American Society for Testing and Materials Philadelphia.

[25] Standard, A., *Standard Test Method for Static Modulus of Elasticity and Poisson's Ratio of Concrete in Compression*, 2010

[26] Hardjito, D. and B. Rangan, *Development and Properties of Low-Calcium Fly Ash-Based Geopolymer Concrete*, 2005, Curtin University of Technology.

[27] Hardjito, D., et al., *On the Development of Fly Ash-Based Geopolymer Concrete.* ACI Materials Journal, 2004. **101**(6)

[28] Junaid, M.T., O. Kayali, and A. Khennane, *Response of Alkali Activated Low Calcium Fly-Ash Based Geopolymer Concrete under Compressive Load at Elevated Temperatures.* Materials and Structures, 2017. **50**(1): p 50

Acknowledgements
The author would like to acknowledge the support provided by the Sustainable Construction Materials and Structural Systems (SCMASS) research group at the University of Sharjah. The research was funded by the research grant number 1602040128-P, provided by the University of Sharjah. The lab work and support provided by Eng Abdul Saboor Karzad, Eng Ahmed Shweiki, Eng Abdul Rehman and Eng Abdallah Albanna is also highly appreciated and acknowledged.

IOP Conf. Series: Materials Science and Engineering **264** (2017) 012005 doi:10.1088/1757-899X/264/1/012005

Architectural membranes on building's functional refurbishment

M Macieira[1], P Mendonça[1] and J Miranda Guedes[2]

1 University of Minho, School of Architecture, Lab2PT, Guimarães, Portugal
2 University of Porto, Faculty of Engineering, Civil Engineering Department, Construct, Porto, Portugal
E-mail: mrfmacieira@gmail.com

Abstract. The aim of this study is to make an insight about architectural membrane applications for building's functional refurbishment, analysing the characteristics that make architectural membranes suitable for this purpose. It is based on sixty interventions used as case studies that are classified chronologically and by countries and uses, pointing out also their applications. Several design strategies are identified and contrasted with the principles set for interventions in existing buildings. The results are illustrated with examples chosen from the case studies, highlighting how membrane building technologies can fulfil the most important principles of building refurbishment.

1. Introduction

Architectural membranes (composite textiles and foils/ films) are used for space defining. Considering its lightness, resilience and flexibility, these are becoming inevitable solutions in specific buildings' refurbishment contexts, when the needs of: weight reduction, enlarging useful floor space area or retrofitting the existing, arise as special requirements. A few millimetres membrane can be a self-supporting material (integrated in a tensile/ pneumatic system) and a selective sunlight filter capable of absorbing or reflecting ultraviolet or infrared light whenever needed; they can also be used to support thermal or acoustic insulation materials [1].

While in the past architectural membranes have been used mainly as a covering system, nowadays a wider range of requirements is emerging and thus a multiplicity of diverse textile/ foil architecture solutions can significantly fulfil many needs dealing with functional aspects to protect and valuing existing buildings. However, despite these benefits, many building practitioners are unfamiliar with the behaviour and the characteristics of membranes. The lack of information about their use and properties, by the design and construction community, limits their capability of achieving the highest possible standards in quality assurance and control on construction projects [1].

2. Interventioned projects sample

To determine how the characteristics of architectural membranes adapt (or fail to adapt) to the demanding requirements of building refurbishment, a detailed analysis of sixty interventions has been conducted. Some of these examples are particularly significant to show the way they achieve integration, or dialogue with the existing building. In order to obtain an overview about this, were prepared data sheets for each intervention project with information about the following aspects: contextual, programmatic, construction, economic and functional. The sixty case studies have been distinguished by the intervention on its building elements: roofs/ courtyards/ skylights (56%), ceilings (25%), facades (12%) and partition walls (7%). All intervention projects have been chronologically classified. According with the durability categories defined by Douglas [2], it is found that most of these refurbishment interventions with membranes occur at old buildings, older than 120 years (37%); considering the other case studies, 35% have a normal life (60-119 years old), 23% have a medium life (30-59 years old), 5% have a short life (under 30 years old) (Figure 1). Membrane's envelope systems have experienced an important development in the last 30 years and an increase in the use of membranes over the last decade is clearly observed, led by applications in roofs/ skylights/ courtyards (Figure 2).

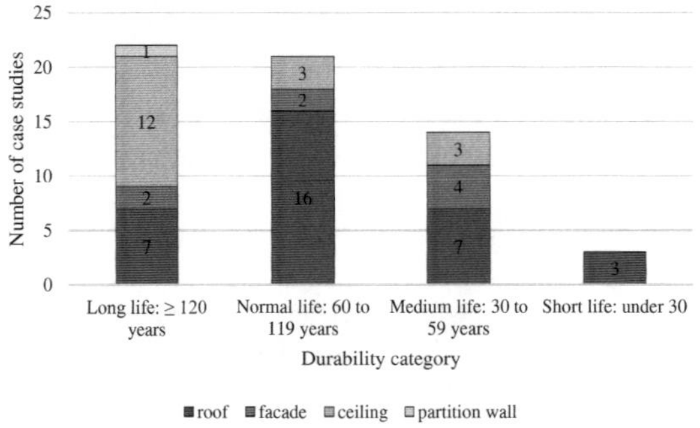

Figure 1. Distribution of the case studies by their durability and interventioned building element.

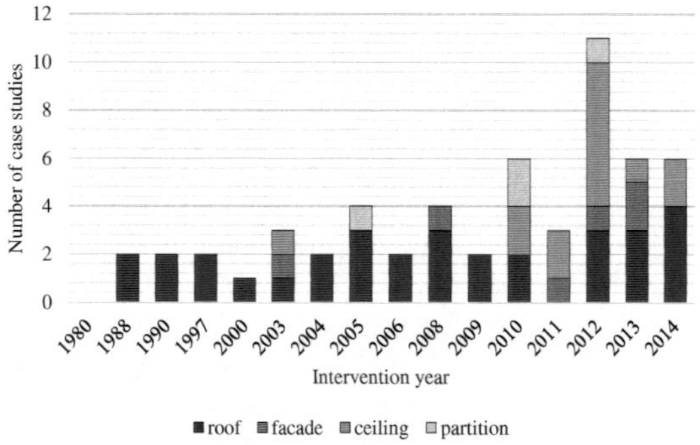

Figure 2. Distribution of the case studies by its refurbishment/ intervention year with membrane building technologies.

Referring to the 19 countries represented, two of them (Germany and UK) top the list with more than 10 interventions (17 – 22%) (Figure 3). Furthermore, Europe accounts with 81% case studies, Asia 10% and America 9%.

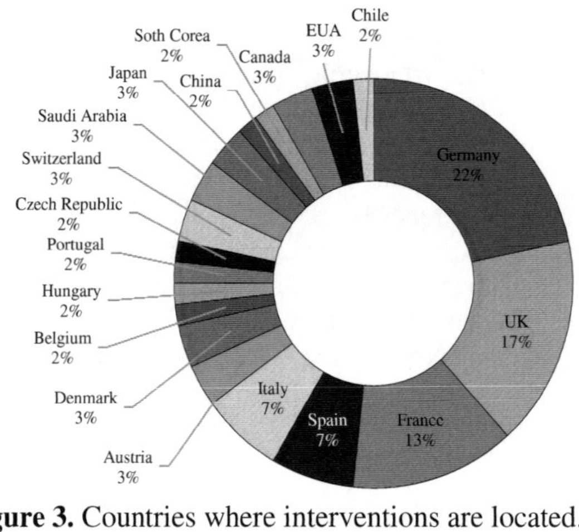

Figure 3. Countries where interventions are located.

The refurbishment of buildings with non-residential use corresponds to 91% of the analysed case studies; cultural (20%), office (17%), sport/ leisure (17%), transport (8%), or religious (11%) uses are those that present more interventions with membranes; buildings with these kind of uses often need to renew its image, or expand its capacity, and membrane materials easily respond to these requirements. Regarding constructive aspects, most of the interventioned buildings presents a mixed structure type (55%) and 90% of existing building elements do not need to be structurally reinforced to accommodate membrane's building technologies. Concerning the building element type, the roof is the most interventioned element with membranes (55%), followed by ceilings (23%). Most intervention projects present tensioned (76%), or pneumatic (24%) building systems. However, most of pneumatic solutions have a particularly domain in roofing intervention projects (where 75% are pneumatic). Most of surface membrane's area of each analysed project range between 51-300 m2 (30%) and 1000-5000 m2 (30%); and mostly used metal profiles as substructure (61%), among other materials such as wood or polymer. In addition, 95% of the analysed intervention projects have fixed membrane installations, being the remaining movable.

2.1. Membrane materials

The use of building technologies that are environmentally more respectful is a growing concern in building design and construction. Textile membranes and films offer an efficient and aesthetic solution not only due to their lightness, mechanical and chemical properties, but also due to the possibility of being functionalized, altering membrane's skin functional properties in a positive way - building's envelope properties can be specifically adapted to changing environmental parameters. New developments in materials, such as low-e coatings, allow membranes to take on more tasks in the area of the building envelope, for example to influence the transmission properties of translucent building skins. Aerogels and phase change materials are in principle suitable to enhance membrane-sandwich structures to become polyvalent components, which can take over advanced tasks with regard to thermal control [3]. Active solar energy use is possible with bendable photovoltaic elements. In the future, other functions, such as electrical conductivity, electroluminescence, or integrated nanostructures will be also relevant [4]. In the field of architectural textile membranes, it is mostly composite textiles that are usually used, consisting of a fabric as carrier and supporting material and a suitable coating to protect and seal the fabric. In the following, basic properties of the most common and state-of-the-art membrane building materials are described on Table 1.

Table 1. Basic common qualities of membrane building materials and corresponding use percentage on analysed case studies.

Polyvinyl chloride (PVC)-coated polyester membrane (16%)	**Polytetrafluoroethylene (PTFE)-coated fiberglass membrane (16%)**	**Ethylene tetrafluoroethylene (ETFE) membrane (35%)**
•less expensive material; •high strength and low weight; •suitable for convertible constructions; •PVC coating emits pollutants when exposed to UV-radiation – limited durability - 15 to 20 years -material gets greyer during the years (due to unfavourable soiling) [5]; •flame retardant.	•increasingly seen as an environmentally friendly and extremely durable - estimated to reach up to 35 years; •fluorine polymer coating acts as anti-adhesive to soiling; •reduced stretch ability and resistance to buckling; •non-combustible.	•used single or in pneumatic cushions; about 0.2 mm thick; •compared to glass, this material offers high UV transmittance [5]; •good coat ability, e.g. with screen printing can be used to limit the heat transmission, or to reflect the solar heat [5]; •stress-strain behaviour and isotropic properties; •rated as not easily flammable.

2.2. Mapping of applications

In order to ensure a good representativeness of the intervention projects sample, it was also necessary to establish a mapping of membrane's application on building's rehabilitation - *Table 2* classifies it into several large families (this list is not exhaustive). From the sample analysis were also identified five basic design options: (1) building inside a building: a membrane structure spans the entire building; (2) creation of facades, skylights and roofs; (3) roofed-over atrium: a courtyard retrofitted with a roof; (4) membrane envelope acting as a second skin; (5) suspended ceilings and stretched partitions.

Table 2. Overlook of main architectural membranes' application on building refurbishment.

Architectural Structures and Spatial definition elements	Floor coverings	Temporary/ auxiliary structures	Structural reinforcement and monitoring	Protection elements	**Installations Devices**
•Tensile/ pneumatic structure, cladding or under layering; •double skin facades; •covering of existing atriums; •horizontal, vertical extension structures; •stretch ceilings and interior partitions (printed, back lighted, micro perforated); •covering for restyling; paint support or as finishing layer.	•Acoustic layer; •waterproofin g layer; •provide safety in the event of a fall.	•Temporary facades/ roof protection; •lost or provisory formwork; •working wired platforms for open spaces; •acoustic and soiling barriers.	•Pneumatic beams; •membrane device for "health structure monitoring" of buildings affected by an hazard; •reinforcement bonding strips.	•Shading and anti UV protection devices; •safety nets; •lightweight thermal insulation composite; •vapour barrier for wood structures.	•Air conducts; •electric extensions (in membrane stripes); •flexible PV integration.

3. Principles of intervention

Most refurbishment interventions in existing buildings are motivated by the fact that functional aspects do not fulfil the necessary requirements, or because the building deterioration endangers their use. These motivations determine the depth of refurbishment interventions, which may lead to significant changes in the buildings. However, it is necessary to prevent those changes to damage the characteristics of the building in an irreversible way.

According to the principles of International Council on Monuments and Sites (ICOMOS) [6], interventions on historic buildings and sites should: (1) preserve their historic character and particular significance; (2) be non-invasive and compatible with the existing size, scale and architectural values; (3) be differentiated from the historic parts and be reversible.

The first principle refers primarily to the replacement's approach - seeking to re-create a missing section or component of the building. This approach requires a high level of accuracy, throughout the design phase. The second principle has to be taken into account mostly in the integration's approach. Adding functions to an ancient building requires additional space, and this leads to a review of the entire structure and to find a dialogue between new and old building components. The last principle has been led by the juxtaposition's approach between old and new elements, where the membrane becomes a light medium between the environments of the past and the present. Thanks to the few following examples, the authors emphasise that the juxtaposition approach seem to be compatible with the outstanding aesthetics of membrane structures, and that the replacement and integration approaches are most often preferred and technically feasible. From the following case studies, some guidance should be concluded for a broader textile application in the future building refurbishment approaches.

In rehabilitation interventions, when new technologies and materials are used to preserve, extend or form some construction, there is always a problem regarding the intrusion's level of new solutions. In any intervention, choosing a solution involves subjective decisions and, therefore, subject to criticism. However intervening and respecting built heritage can be compatible actions; each structure is unique and only a careful study enables to select the intervention that best suits each situation [7].

3.1. Replacement – compatibility

Seeking to re-create a missing section or component of a building, this approach requires the highest level of accuracy throughout the design phase, in order to: (a) preserve the character of the existing building; (b) avoid any future problems, guaranteeing the compatibility of new elements with the existing building. Regarding compatibility aspect, this approach considers that the intervention should be compatible with the existing size, scale and architectural values.

Diagnosis is a very important design tool, because it allows selecting the best rehabilitation design options [7]. For example, in the intervention project on Dresden's train station [8] (Figure 4a) the option to replace the existing roof by a membrane was taken considering the structure's condition. Due to membrane's lightness it was possible to maximize the preservation of existing structural elements, improving, at the same time, the indoor natural lighting. Another example is the intervention project Camellia house [9] (Figure 4b) – one of the Europe's oldest cast iron structure. An inspection to the existing building revealed that the cast iron structure was debilitated to withstand a variety of overloads, such as snow or wind, and that there were many broken windows due to vandalism acts. It was also found that the existing glass roof, which did not withstand snow and wind loads, had a reduced slope. Thus, to protect the structure, it was designed a second roof with ETFE membrane (with slope correction) that fits over the existing glass/iron roof. The ETFE membrane, applied with aluminium profiles, is sufficiently resistant to withstand snow loads and don't break - making it safer than glass. Therefore the building can maintain its function, as the transparent ETFE membrane ensures: necessary light transmission; reduced overload to existing structure and opening panel's incorporation to allow the roof's maintenance. One of the difficulties of this project was the ETFE membrane's roof modelling due to geometric complexity of existing roof. Figure 5 shows other intervention projects where functional rehabilitation was made through a replacement approach / recreation of missing parts with systems that integrate membrane materials.

Figure 4. (a) Dresden's train station (before and after) [8]; (b) Camellia house (before and after) [9].

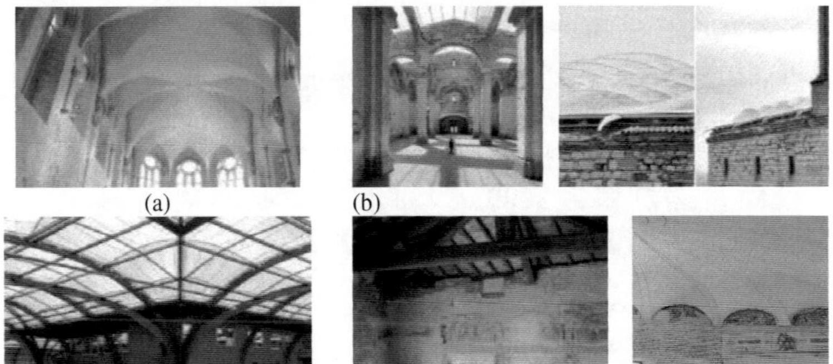

Figure 5. (a) Kostel Povýšení sv. Kříže [10] – reconstitution of vaulted ceiling to achieve higher acoustic performance; (b) Corbera d'Ebre church [11]; (c) Ténnis club bourg-la-reine [12] – roof replacement to increment natural lighting performance; (d) Bologna museum [13].

3.2. Integration – adaptability

Adding functions to an ancient building requires additional space, and this leads to a review of the entire structure to find a dialogue between new and old materials and techniques. It becomes a meticulous work with the objective of finding: (a) linguistic coherence; (b) structural compatibility and (c) feasible connexion.

When intervening in existing buildings, usually composed by orthogonal planes, membrane materials have to put aside their morphological freedom. This requires the design of transition components that ensure the correct connection between membranes and existing building. This integration is enabled by the lightness and other physical properties of the membranes (i.e. which do not overload the existing building, while providing protection and functional performance improvements).
In the recovery and reuse of urban fabric, membranes can "repair" dismembered parts, giving new identity and functions to secondary spaces. This is the case of "Imagination Headquarters" (UK, 1990) [14] (figure 6a) – a rearrangement of an existing space between two buildings, eliminating a number of service facilities between them, to create a large internal covered gallery with a PES/ PVC membrane (surface area of approx. 590m2) - that ensures necessary light conditions to host a new use/ function. Another example is the Wismar Nursery [15] (figure 6b), where the enclosing space between two existing volumes created a buffer zone that increase the thermal regulation and ventilation of adjacent spaces. The atrium can be shaded through the air pressure adjustment of the pneumatic roof with triple ETFE membrane. Thus, it was not necessary to refurbish the facades that face the created atrium, so its thermal inertia remained unchanged.
Fukiya parish council [16] presents a room extension, aiming to build a large meeting room space without intermediate supports. The connection between new and old was inevitable; the new roof had to be compatible with the structural configuration and architectural characteristics of the surrounding buildings. Naturally, the first option was a tiled roof; however the users desired a copper roof, which was an expensive option. Hearby, as alternative, it was chosen a glass fibre/ PTFE membrane dyed in a similar colour of the surrounding ceramic roofing tiles (figure 6c), not only due to economic factors, but also to the possibility to achieve a bigger span without intermediate supports.
Originally built in 1900, the interventioned project Logan Offices [17] (figure 6d) resulted in the conversion of an industrial warehouse into offices. To adapt/ convert this space it was necessary to create new compartments and adapt it to acoustic and luminous requirements for the new use. Thus, it was decided to add a second interior skin, in a translucent membrane, to the existent vertical and horizontal surfaces. The suspended ceilings, with translucent and back-lit membrane, enable artificial light diffusion without shadows' creation. The vertical lining, with a stretched PES/PVC open mesh membrane, acts as an unifying element of intervention, hides any aesthetic imperfections, keeps the passage of natural light, improves acoustic performance (absorption) and keeps the viewing of original/ existing building elements (such as frames, pipes and iron structure).

(c)

(a) (b) (d)

Figure 6. (a) Imagination headquarters - exterior and interior views [14]; (b) Wismar nursery - before and after [15]; (c) Fukiya parish council [16]; (d) Logan office – interior view.

3.3. Juxtaposition – durability, repeatability and reversibility
Little guidance can be generally assumed from the juxtaposition approach, which requires a different strategy every time, finding the right balance between two opposite requirements: (a) how to differ the new elements from the old ones and (b) how to minimize the visual impact of new buildings in order to guarantee the perception of pre-existing elements. With this approach the intervention aims to be

differentiated and reversible from the existing parts - reversibility of the new intervention is the most important aspect that designers have to take into account. The durability requirement of refurbishment solutions for old buildings with historic and cultural value is more severe than in the case of current modern buildings, because its useful lifetime is significantly bigger. In consequence, regarding a repeatability approach, it is possible to consider that, depending on interventions' specifications, the use of membranes, even with a durability of 10-35 years, may be viable solutions. Attending to constant functional changes that a building is called to support nowadays, whether for social and economic, among other reasons, it requires flexibility of use, and membrane building technologies can meet these needs, extending building's useful life. The use of membranes can also be include into the group of Reversible design, due to the fact of being lightweight solutions, easily mounted/ dismounted and with a reuse/ recycling process relatively simple – characteristics that were presented in all intervention projects analysed.

Above all, it may be considered that refurbishment solutions with membrane are repeatable, i.e. when they reach the end of their useful life, they can be re-implanted. In the intervention project on a 19th century house "Evidencia light hotel" [18] (Figure 7a), a membrane lining solution was chosen due to the reduced room dimensions and the possibility of renew/ reuse it in future scenarios. In addition, it allowed to: achieve useful area gains (due to its reduced thickness), include acoustic insulation, hide existent surfaces' imperfections and constitute a substrate that could be printed.

3.4.Roof-over

Membranes are increasingly replacing glass in roofing structures on the retrofit of buildings with courtyards. Here, the advantages of membranes are: lightness; the possibility of having large surfaces with a single membrane (5mx20m) (which requires fewer structural supporting elements) and its satisfactory durability. In addition, in most of the interventioned projects, the connections to the existing buildings is reversible, causing no significant damage to it.

In the case of the intervention project Irgatza palace [19] (Figure 7b), classified as cultural built heritage, a pneumatic ETFE membrane roof was applied over the existing atrium to preserve the building (with wood structure) from the effects of climatic changes, mainly from rain (very frequent in this local), improving its hygrothermal performance and expanding its usefull floor area at the same time. At Fuchun Kosa Zou Ma Lou [20] (Figure 7c) - a vernacular courtyard house - the existing courtyard was covered with a single transparent PVC membrane, supported by a bamboo structure, to increase the useful floor area. The membrane can be partially open to promote space ventilation and thermal regulation and its configuration promote rainwater collection for later reuse.

The intervention project "Shishiodoshi house" [21] (Figure 7d) resulted in a vertical extension of a building built in 1980. With a height of eleven meters, this vertical extension was made with a wooden structure totally covered with a stretched white colour PES/PVC membrane. The membrane was chosen due to its lightness and reduced cost.

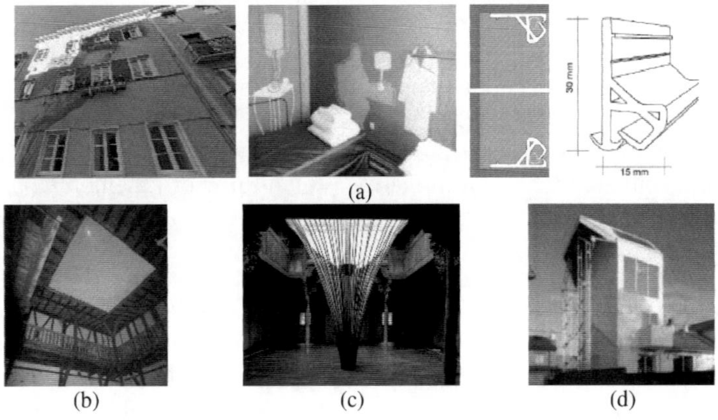

Figure 7. (a) Evidencia light hotel [18]; (b) Irgatza palace [19]; (c) Fuchun Kosa Zou Ma Lou [20]; (d) Shishiodoshi house [21].

IOP Conf. Series: Materials Science and Engineering **264** (2017) 012005 doi:10.1088/1757-899X/264/1/012005

4. Conclusion

Results analysis highlights an accurate walk-through architectural membrane performance, in accordance with design refurbishment principles.

The case studies' sample shows the wide range of possible uses for architectural membranes in refurbishment projects - highlighting good practices in their current application and defining guidelines to support decision-making in future.

Textile/ foil materials present inherent qualities and performances that motivate the use of membranes in building refurbishment, such as: resistance; structural reinforcement and flexibility; adaptability; lightweight; ease of implementation; durability and weather change resistance; electromagnetic/ electrostatic/ anti corrosion; passive of being functionalized; reduced economic and environmental impact and architectural aesthetic - allowing design of reversible and repeatable solutions. Nevertheless, the performance of these materials in refurbishment interventions, as well as the design methodologies need to be further investigated in order to assure the better response to each specific context – in a close relation with the peculiarity of existing buildings.

5. References

[1] Macieira M Mendonça P and Guedes J 2016 *Functional Rehabilitation of Built Industrial Heritage with Architectural Membranes: Contextual, Climatic and Functional Aspects* (Lisbon: proceedings of *III International Congress on Industrial Heritage – Reuse of Industrial Sites: a Challenge for Heritage Conservation*) p. 169, 2016

[2] Douglas J 2009 *Building Adaptation* (UK: Elsevier) Heriot-Watt University, Edinburgh.

[3] Cremers J and Lausch F 2008 *High-Performance Insulation Made of Translucent Silica Aerogels for Membranes* (Germany: DETAIL magazine) issue 5-2008 Plastics, pp. 524-530.

[4] Cremers J 2007 *Flexible Photovoltaics Integrated in Transparent Membrane and Pneumatic Foil Constructions* (Lausanne: Proceedings Conference CISBAT, EPFL).

[5] Knippers J et al. 2011 *Construction Manual for Polymers and Membranes* (Germany: Detail & Birkäuser) p. 217.

[6] ICOMOS 2011 *The Paris Declaration on Heritage as a Driver of Development* [Online]. Available: http://www.international.icomos.org/Paris2011/GA2011_Declaration_de_Paris_EN_20120109.pdf

[7] Guedes J et al. 2003 *Conservação e Reforço de Estruturas Património* (in portuguese) (Cascais: Jornadas Europeias do Património).

[8] Foster + Partners 2016. *Desden Train Station Redevelopment* [Online]. Available at 08 June 2017: http://www.fosterandpartners.com/projects/dresden-station-redevelopment/

[9] Architen 2016 *Camellia House* [Online]. Available at 08 June 2017: http://www.architen.com/projects/camellia-house

[10] TIC 2016 *Jihlava* [Online]. Available at 08 June 2017: http://tic.jihlava.cz/ES/vismo/o_utvar.asp?id_org=101078&id_u=43239.

[11] Designboom 2016 *Ferran Vizioso Frames a Derelict Church* [Online]. Available at 08 june 2017: http://www.designboom.com/architecture/ferran-vizoso-architecture-frames-a-derelict-church

[12] Serge Ferrari 2016. *Uma Renovação Luminosa* [Online]. Available at 08 June 2017: http://pt-br.sergeferrari.com/arquitetura-leve/uma-renovacao-luminosa-membrana-precontraint-1202-s2/.

[13] Barrisol 2016 *Church Projects* [Online]. Available at 08 June 2017: http://www.barrisolsydney.com.au/our-projects/church-projects/

[14] Architen 2016 *Imagination Headquarters* [Online]. Available at 08 June 2017: http://www.architen.com/projects/imagination-headquarters/;

[15] ENOB 2016 *Children day care center, Wismar.* [Online]. Available at 08 June 2017: http://www.enob.info/en/refurbishment/projects/details/childrens-day-care-centre-in-wismar/.

[16] Ishii K 1995 *Membrane Structures in Japa* (Tokyo: SPS publishing company) pp.307 -309.

[17] Newmat USA 2016 *Logan Office* [Online]. Available at 08 June 2017: http://newmatusa.com/newmat/2400/projects/corporate/logan.html.

[18] Evidencia 2016 *Evidencia Light Hotel* [Online]. Available at 08 June 2017: http://evidencia-light-sta-catarina.lisbon-hotel.net/pt/.
[19] Bat Spain 2016 *Palacio Igartza* [Online]. Available at 08 June 2017: http://www.batspain.com/etfe.html.
[20] Archimixing 2016 *Fuchun Kosa Zou Ma Lou* [Online]. Available at 08 June 2017: http://www.archmixing.com/index.php/Works/shownews/id/363.
[21] Avignon clouet 2015 *Un Batiment, Combient De Vies?* [Online]. Available at 08 June 2017: http://www.avignon-clouet.com/home.php/?p=846

Acknowledgments
First author thanks to FCT (Fundação para a Ciência e Tecnologia – Portugal), MCE (Ministério da Educação e Ciência – Portugal) and ESF (European Social Fund) for supporting the research fellowship with the reference SFRH/BD/104891/2014; and with other authors thanks the financial support from the Project UID/AUR/04509/2013 by FCTMEC by national funding and, when applicable, FEDER co-financing under the new PT2020 partnership agreement - Lab2PT, School of Architecture/ University of Minho, Portugal and Project POCI-01-0145-FEDER-007457 - CONSTRUCT - Institute of R&D In Structures and Construction of Faculty of Engineering/ University of Porto, Portugal, funded by FEDER funds through COMPETE2020.

IOP Conf. Series: Materials Science and Engineering **264** (2017) 012006 doi:10.1088/1757-899X/264/1/012006

Assessment on bamboo scrimber as a substitute for timber in building envelope in tropical and humid subtropical climate zones - part 1 hygrothermal properties test

Zujian HUANG[1,2], Yimin SUN[1] and Florian MUSSO[2]

1 School of architecture, State Key Laboratory of Subtropical Building Science, South China University of Technology, Guangzhou 510640, China
2 Chair of Building Construction and Material Science, Department of Architecture, Technical University of Munich, Munich 80333, Germany
E-mail: arymsun@scut.edu.cn

Abstract. Bamboo scrimber was bamboo fiber based panel developed in 2000s that was potential to be an ideal substitute for timber in bamboo growing areas. For obtaining material parameters and evaluating the performance in building envelope, bamboo scrimber was systematically tested for hygrothermal properties, based on the building envelope heat and moisture process model. Static test items included density calculation and vacuum saturation test for basic properties; sorption test for moisture storage properties; capillary absorption test, water vapour transmission test and drying test for moisture transport properties; thermal analysis for heat storage properties; thermal conductivity test, surface light and thermal properties test for heat transport properties. The test results, by comparison with reference timbers showed that bamboo scrimber had higher heat storage and heat transport properties and lower moisture storage and transport properties. The dynamic test in wind tunnel with outdoor weather condition showed that bamboo scrimber had lower moisture absorption and desorption rate than reference hardwood. The significant magnitude difference between the static and dynamic test results showed the necessity of a comprehensive evaluation approach that could take more practical conditions into consideration.

1. Introduction

Bamboo forests were widely distributed in the tropical and humid subtropical climate zones in the Asia Pacific (ca. 67%), the Americas (ca. 30%) and Africa (ca. 3%) [1]. The application of bamboo in building industry varied with regional differences due to the local forest resources, economic conditions and construction technology. Large-diameter round bamboos, as an easily accessible, affordable and seismic construction material, were widely used in traditional bamboo building in the Southeast Asia and South America [2]. Since 1970s, restricted by the regional wood forest resources, modified bamboos in standard panel or square form were considered to be an ideal substitute for timber [3]. In countries including China, India, Thailand, Vietnam, Costa Rica, Malaysia, etc. timber processing technologies were introduced to the industrial utilization of bamboo, afterwards plybamboo (1980s), bamboo particleboard and bamboo OSB (1990s), bamboo laminated lumber (1990s) were successively developed [4,5,6], and promoted to the concrete formworks, load-bearing components, truck and bus bottom boards, furniture and finishes industries [3].

Microscopically, bamboo was composed of mostly parallel arranged fiber that had higher strength than wood fiber [1,2], which prompted that products based on bamboo fiber were beneficial for exploiting the inherent advantage. However, due to the low added value and complicated manufacturing process,

bamboo fiber boards in form of MDF consisted of fully separated fiber weren't successfully promoted [7]. Bamboo scrimber simplified the constituent units as loose bamboo fiber bundle[8], could be made from those herbaceous and small dimension bamboos, and improve the raw material utilization rate to 90% [9], therefore rapidly grew as the mainstream of bamboo fiber boards and was widely promoted to the load-bearing components, indoor finishes, outdoor pavement, furniture and wind power blades, etc.[9,10] existing study on bamboo scrimber covered the manufacturing process, including the produce of constituent unit, dipping, assembly, cold and hot pressing method[11,12] and cracking proofing, deformation proofing, anti-corrosion [13] and frame retardant techniques [14].
In practical application, due to the insufficiency of material study, bamboo properties were commonly substituted with timber parameters, which caused inaccuracy to the projects. The paper tested bamboo scrimber for the hygrothermal properties with static and systematic standard methods, and afterwards carried out a wind tunnel test to examine the hygric behavior in outdoor condition. Corresponding timbers in both tests were set as reference for the evaluation of bamboo scrimber.

2. Hygrothermal properties test

2.1. Material

Before the test, based on the investigation of literature, market and material producers in bamboo industry, typical bamboo scrimber samples were chosen as the test objects, as shown in Table 1.

Table 1. Information of the bamboo scrimber sample.

BFB (Bamboo scrimber)	Constituent unit	Assembly	Thickness	Main application	Sample sources
	Bamboo bundle (width:10-30mm)	Parallel	30.0 mm	Load-bearing component, indoor/outdoor flooring, finishes, furniture	Dasso Industrial Group Co. Ltd. Hangzhou, China

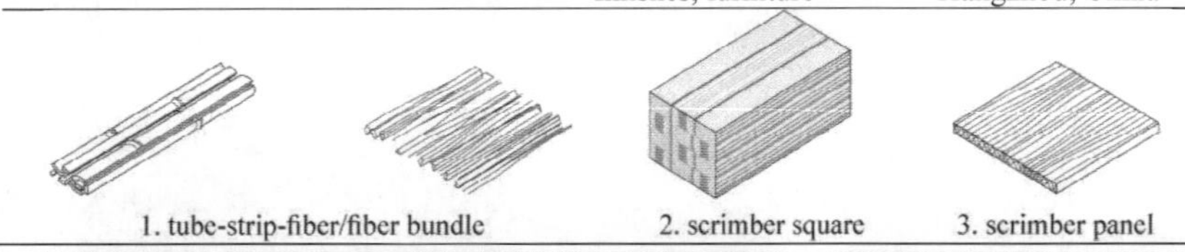

Main manufacturing processes of bamboo scrimber (Photos were taken in Zhejiang province, China by the author)

2.2. Static test items

There were dozens of hygrothermal properties that constituted different heat and moisture process models for simulation. Before 1990s, the steady vapor permeation model proposed by Glaser was widely used, which took purely the water vapor permeability for the calculation in one-dimensional steady-state moisture process, which couldn't describe the actual situation that was generally non-steady and multiphase simultaneous. In the nearly 20 years, Molend, Pedersen, Künzel, Häupl, Janssen, Mendes, Steeman and Tariku successively proposed heat-air-moisture transfer model (HAM model) to describe the coupled storage and transport of heat, moisture and air in building envelope [15,16,17,18,19,20,21,22] The models had been verified under various conditions, including the IEA Annex 41 in 2008 that compared 17 HAM models and their numerical simulation tools and completed 7 joint comparative tasks, which proved that the calculation results among the main HAM models and the measured results were close to each other [23,24]. HAM models normally consisted of highly coupled nonlinear partial differential equations that describe simultaneously the heat and moisture processes in building envelope, which were solved through numerical methods [25] Compared with the Glaser model, HAM models relied on much more detailed material parameters as calculation inputs. Hartwig M. Künzel simplified the necessary material parameters for one and two-dimensional heat and moisture process calculation, and used vapor pressure and relative humidity respectively as the driving potentials for gaseous water and liquid water. Two coupled differential equations were proposed to describe the non-steady heat and moisture transport processes in building components, which formed the basis for the computer program WUFI [17].

The heat and moisture transport mechanism and test items were showed in Figure 1, which were based on the equations from H M. Künzel. The above equation described the moisture process, which successively consisted of moisture storage, liquid water transport and gaseous water transport. The below equation was the heat process, which was successively composed of heat storage, heat transport and heat sink. Both equations consisted of storage terms on left side and transport terms on right side.

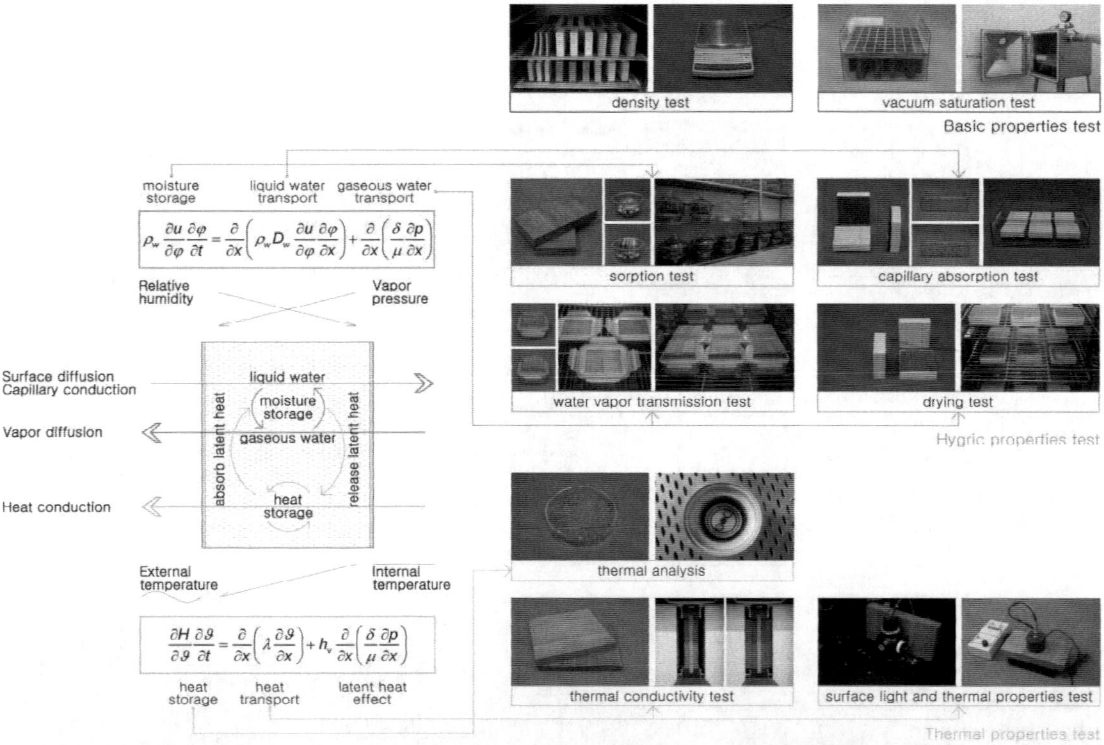

Figure 1. Building envelope heat and moisture transport mechanism and the bamboo scrimber test items (the equations on left side were from H M. Künzel [17]).

The running of the model based on these coupled differential equations required the input of material parameters, climate data, internal conditions, and control parameters such as numerical grid and time step width. The material parameters included basic properties such as bulk density, porosity, and the correlative properties characterized the storage and transport abilities of moisture and heat. Nine test items were carried out according to the corresponding test standards. The tests targeted only for the thickness direction, considering that the thickness of the bamboo panels in practical application was far less than the plane dimensions, which meant the heat and moisture transferred approximately one-dimensionally and vertically through the panels. As some previous studies had shown that the repeatability error was generally far smaller than the unevenness error of the samples [26], multiple copies were carried out simultaneously during the test, rather than repeated the specimens for the same items. (Table 2)

Table 2. Operation method, equipment arrangement and specimens treatment of the test.

Category	Item	Operation method	Equipment arrangement	Specimens treatment
Basic properties	1. Density test Target value: dry bulk density	-	Drying oven: digital stainless steel electric blast oven 101A-2S, accuracy ±1℃ Balance: SHIMADZU UX6200H, accuracy 0.01g	Quantity: 9 copies *a Size: 10×20cm
	2. Vacuum saturation test Target value: open porosity, vacuum saturation moisture content	Refer to the American standard ASTM D7370-2009, and the European standard DIN EN 1936-2007 [27, 28]	Drying oven: same as item 1 Balance: same as item 1 Vacuum chamber & pump: Sliding vane rotary vacuum pump Fopump-2X-8, speed 8L/s, Extreme pressure 600Pa	Quantity: 3 copies Size: 10×10cm
Hygric properties	3. Sorption test Target value: isothermal absorption and desorption curve	Refer to the international standard ISO 12571:2012 and the American standard ASTM C1498-04a [29, 30]	Vacuum dryer (9 copies): inner diameter= 30cm Balance: same as item 1 T & RH recorder: TH10R-EX, with external sensor (9 copies), accuracy T±0.2℃, RH±2%	Quantity: 3 copies, including 1 large sample and 2 small samples Size: large sample 10×20cm *b, small samples 4×5cm
	4. Water vapor transmission test Target value: water vapor transfer coefficient	Refer to the international standard ISO 12572:2001(E) [31]	Dry cups (3 copies), wet cups (3 copies) Balance: same as item 1 Constant T&RH curing box: HWS-250B, accuracy T±0.5℃, RH±5%	Quantity: 6 copies Size: 10×10cm
	5. Capillary absorption test Target value: water absorption coefficient; capillary saturation moisture content	Refer to the international standard ISO 15148:2002(E) [32]. With the results of water absorption coefficient and capillary saturation moisture content, liquid transport coefficient was then calculated: [33] $D_l=\pi/4\cdot(A_{cap}/w_{cap})^2$	Water sink: length×width×height= 40×30×10 cm; Stainless steel nuts supports of equal height were fixed on the bottom; Distilled water were injected in the sink and kept the water level 3mm above the supports Balance: same as item 1	Quantity: 3 copies, sealed on top and side surfaces, while kept 4 5×5mm air vents on top surface and 6mm gaps close to the bottom edges on side surfaces open
	6. Drying test Target value: drying curve; drying rate curve	Refer to the test method of Fraunhofer IBP (accredited according to DIN EN ISO/IEC 17025)	Balance: same as item 1 Constant T&RH curing box: same as item 5	Quantity: 3 copies, sealed on top and side surfaces Size: 10×10cm *c
Thermal properties	7. Thermal analysis Target value: specific heat capacity	Refer to the international standard ISO 11357-4-2005 [34]	Differential thermal scanning / thermogravimetric analyzer: TA4000/ 2910MDSC Reference material: sapphire	Quantity: 3 copies Mass: 20mg
	8. Thermal conductivity test Target value: thermal conductivity	Refer to the international standard ISO 11357-4-2005 [35] 24 hours heat storage coefficient was calculated by: $S_{24h}=0.51\cdot(\rho\cdot c\cdot\lambda)^{0.5}$	Guarded hot plate apparatus: CD-DR3030, hot plate T=35℃, cold plate T=15℃, size of central heat transfer area 15×15cm, accuracy ±2%	Quantity: 4 copies, including 3 large samples and 1 small sample Size: large samples 30×30cm, small sample 10×20cm *d
	9. Surface light and thermal properties test Target value: hemispherical emissivity; light reflectivity; solar direct reflectivity; solar direct absorptivity	Refer to the European standard DIN EN 16012-2012 [36]	D&S Digital hemisphere emissivity tester: AE1/RD1, Linear relationship ± 0.01units Ultraviolet, visible, near infrared spectrophotometer U-4100: measurement wavelength range 200-2600nm	Quantity: 4 copies, including 1 large sample and 3 small samples Size: large sample 10×20cm *d, small samples 10×10cm

*a. Specimens after being dried in item 1 were put in the vacuum dryers for test item 3;
*b. Specimens after reaching the equilibrium moisture content were carried on in test item 8 and 9;
*c. Use the specimens after vacuum saturation test in test item 2;
*d. Use the specimens after reaching the equilibrium moisture content in test item 3

2.3. Dynamic test in wind tunnel

For examining the behavior of the materials in outdoor conditions, a wind tunnel test was carried out with a typical summer day weather data of Guangzhou, a subtropical city located in south China. The weather data including dynamic solar radiation, air temperature, relative humidity and constant wind speed was repeated for 72 hours, of which the last 48 hours were selected for analysis. The reference timber in this test was Intsia.spp (Caesalpiniaceae), a kind of antiseptic hardwood for outdoor flooring. (Figure 2, Table 3)

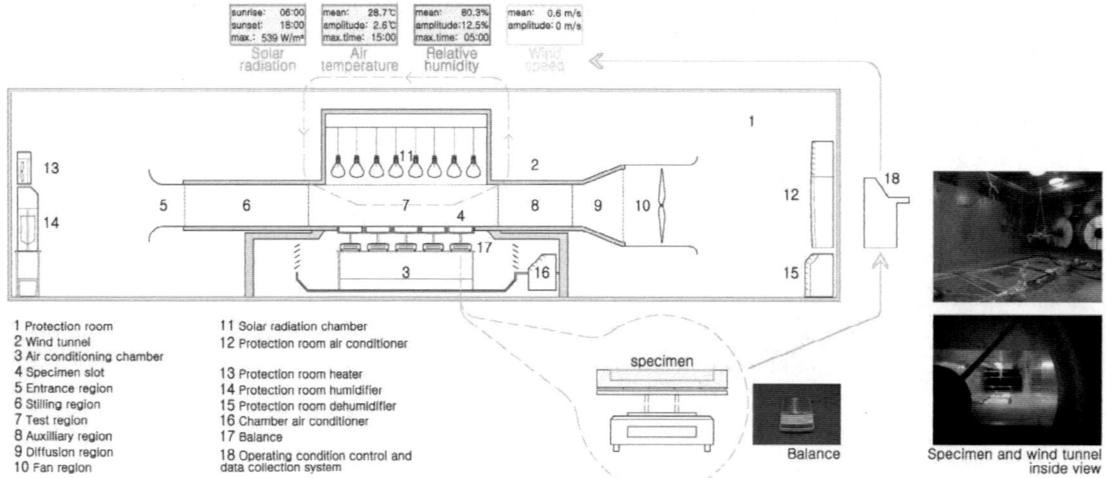

Figure 2. Bamboo scrimber wind tunnel test.

Table 3. Weather condition, operation, equipment and specimens of the wind tunnel test.

Weather condition (24h repeated)	Operation method	Main equipment arrangement	Specimens treatment
Solar radiation: sunrise - 06:00 sunset - 18:00 max. value - 539.0W/m^2 Air temperature: mean value - 28.7℃ amplitude - 2.6℃ max. time - 15:00 Relative humidity: mean value - 80.3% amplitude - 12.5% max. time - 05:00 Wind speed: constant as 0.6m/s	Operated the weather condition until deviation of solar radiation, and relative humidity and wind speed ≤5%, and air temperature ≤0.3℃; Installed the specimens with slots at 06:00 and started mass record by electronic balance (range 0-4kg, accuracy 0.01g; time step: 10s); Hourly mass values m_i were calculated from the average values of all the instantaneous measured values within the hour; The hourly absorption and desorption rate $U_{ad,i}$ value was calculated from: $U_{ad,i} = (m_i - m_{i-1})/A$	HHCWT: Hot-Humid Climatic Wind Tunnel, including control systems for: Air temperature: 2 air conditioner (KFRd-60LW/VA-ZXF, 6kW, and KFR-72LW/08FZBPC-a, 7.2kW) and 2 electric heating fans (1.5kW and 5kW), range 20-40℃, accuracy 0.3℃; Relative humidity: 2 electrode humidifiers (BFD-01-04, 0.4kg/h) and 1 dehumidifier (DH-890C, 90L/d), range 40-90%, accuracy 3%; Solar radiation: 8 infrared lights (305 - 3000nm), range 0-1030W/m^2, accuracy 10W/m^2; Wind: axial fan, range 0-5m/s, accuracy 0.2m/s.	Size: A_{BFB} - 0.0822 m^2 (30×27.4cm) A_{HW} - 0.0810m^2 (30×27cm) Side and bottom surfaces sealed with 1-2mm Vaseline The gaps between specimens and the slots were also sealed with Vaseline

3. Test results analysis

3.1. Reference timber

Since the knowledge on timber was relatively sufficient, the comparison with timber could support the evaluation of bamboo scrimber. However, the timber parameters from different databases varied greatly, which meant that to obtain precise Bamboo-Timber ratio was impractical. Here the 28 timbers from Fraunhofer IBP in WUFI Plus material database were chosen as reference timber (RT). The range of RT defined by the maximum and minimum values, and the relative position of bamboo scrimber were analyzed below. (Table 4, Figure 3)

3.2. Comparison between bamboo scrimber and the reference timber

1) Basic properties. Both the bulk density and open porosity of bamboo scrimber exceeded the range of the reference timber. The high bulk density, as 1108.77 kg/m^3, rarely existed in plant materials,

which mainly due to the high density of the raw material and strong pressing during the manufacturing process. The low open porosity resulted from the high pressure and waterproofing agent that compressed and blocked the pore structure, was conductive to the application in outdoor environment.
2) Hygric properties. Compared with the reference timber, the isothermal absorption and desorption curve of bamboo scrimber was in a lower position. The gaseous water transport property was greatly influenced by the moisture content. The vapor transfer resistance factor fell from 921.28 to 47.67 when moisture content rose from $w_{RH=20.0\%}$ to $w_{RH=93.0\%}$. The relative position of bamboo scrimber to reference timber also fell with the increasing moisture content. Due to the lower open porosity, the liquid water storage and transport abilities were lower than the reference timber.
3) Thermal properties. Both the heat storage and transport properties of bamboo scrimber were higher than the reference timber, which mainly caused by the high density. The thermal conductivity was in nonlinear relation with the moisture content, and that was normally described by an approximately linear fitted thermal conductivity supplement value, which was 0.0984 [%/M.-%] in this test.

Table 4 Bamboo scrimber hygrothermal properties test results and the comparison with RT.

	Items	Unit	BFB -Bamboo scrimber		Reference Timber Min. value		Reference Timber Max. value	
Basic properties	Dry bulk density ρ	[kg/m^3]		1108.77		400		708
	Open porosity Φ	[-]		17.36%		41%		90%
Hygric properties	20℃ isothermal absorption and desorption curve	[kg/m^3]	$w_{RH=11.2\%}$	6.79	$w_{RH=0\%}$	0	$w_{RH=0\%}$	0
			$w_{RH=24.4\%}$	14.05	$w_{RH=20\%}$	5.70	$w_{RH=20\%}$	42.00
			$w_{RH=33.4\%}$	20.26	$w_{RH=30\%}$	9.70	$w_{RH=30\%}$	61.00
			$w_{RH=43.5\%}$	25.55	$w_{RH=50\%}$	22.10	$w_{RH=50\%}$	74.00
			$w_{RH=55.0\%}$	30.19	$w_{RH=65\%}$	31.90	$w_{RH=65\%}$	90.00
			$w_{RH=59.7\%}$	32.24	$w_{RH=80\%}$	45.50	$w_{RH=80\%}$	115.00
			$w_{RH=77.2\%}$	53.75	$w_{RH=90\%}$	62.29	$w_{RH=90\%}$	151.00
			$w_{RH=85.4\%}$	69.58	$w_{RH=93\%}$	80.10	$w_{RH=93\%}$	197.00
			$w_{RH=96.3\%}$	155.04	$w_{RH=95\%}$	88.00	$w_{RH=95\%}$	245.00
					$w_{RH=97\%}$	94.40	$w_{RH=97\%}$	318.00
	Water vapor transfer coefficient ($\mu_{RH=0.0\%}$ and $\mu_{RH=100.0\%}$ were regarded as equal respectively to$\mu_{RH=20.0\%}$ and $\mu_{RH=93.0\%}$)	[-]	$\mu_{RH=20.0\%}$	921.28	$\mu_{RH=0.0\%}$	9	$w_{RH=0\%}$	845
			$\mu_{RH=25.0\%}$	781.24	$\mu_{RH=100\%}$	7	$w_{RH=25\%}$	845
			$\mu_{RH=35.0\%}$	600.96			$w_{RH=75\%}$	376
			$\mu_{RH=45.0\%}$	489.50			$w_{RH=100\%}$	376
			$\mu_{RH=50.0\%}$	446.94				
			$\mu_{RH=73.0\%}$	106.15				
			$\mu_{RH=83.0\%}$	73.79				
			$\mu_{RH=93.0\%}$	47.67				
	Drying rate (T =23℃, RH=50%)	[E-07kg/(m^2·s)]	$U_{u12\%-u11\%}$	12.58				
			$U_{u11\%-u10\%}$	11.46				
			$U_{u10\%-u9\%}$	10.38		-		-
			$U_{u9\%-u8\%}$	9.71				
			$U_{u8\%-7\%}$	8.98				
			$U_{u7\%-u6\%}$	7.74				
	Capillary saturation moisture content w_{cap}	[kg/m^3]		165.93		326.00		864.50
	Water absorption coefficient A_{cap}	[kg/(m^2·s$^{0.5}$)]		0.000873		0.0004		0.0045
Thermal properties	Specific heat capacity c	[J/kg·K]		1550		1300		2100
	Thermal conductivity λ_{dry}	[W/m·K]		0.1625		0.09		0.13
	24h heat storage coefficient S_{24h}	[W/m^2·K]		8.52		3.62		5.55
	Hemispherical emissivity ε	[-]		0.66				
	Light reflectivity (380-780nm) ρ_v	[%]		19.74				
	Solar direct reflectivity (200-2600nm) ρ_e	[%]		44.51		-		-
	Solar direct absorptivity (200-2600nm) α_e	[%]		55.49				

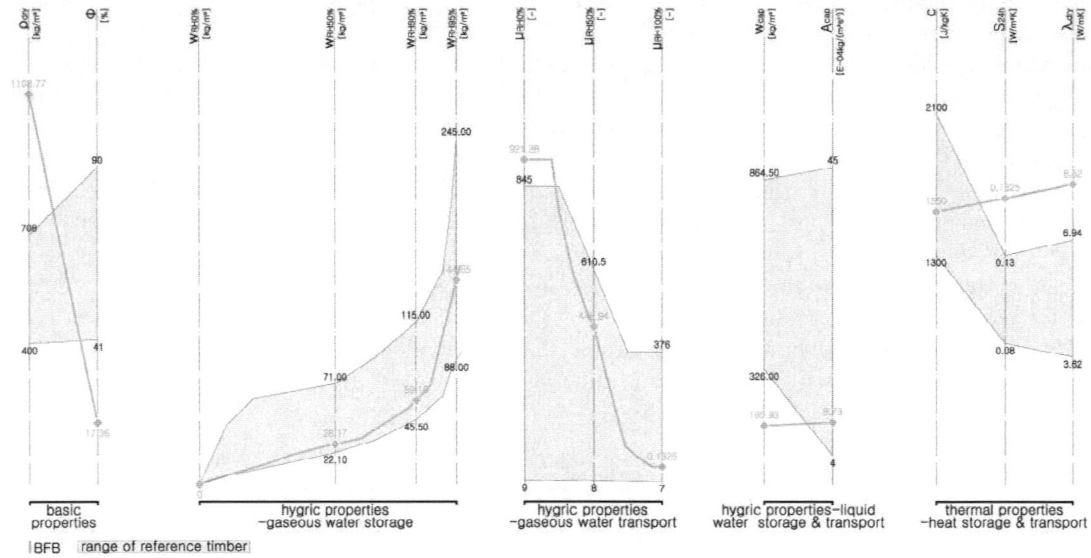

Figure 3. Relative position of bamboo scrimber parameters in the range of reference timber.

3.3. Wind tunnel test results

1) The hourly mass change rate comparison between bamboo scrimber and hardwood showed that bamboo scrimber had slower absorption and desorption rate U_{ad} value. Except certain areas, mostly the absorption-desorption transition stage, the hourly U_{ad} values of bamboo scrimber were up to 33.69 E-06 kg/m^{2}s smaller than that of hardwood specimen. (Figure 4)

2) Affected by the strong outdoor solar radiation and temperature, the absorption and desorption rate of bamboo scrimber were respectively up to 50.22 E-06 kg/m^{2}s and 67.41 E-06 kg/m^{2}s, much higher than the values 7.74-12.58 E-07 kg/m^{2}s resulted from the static drying test (T =23℃, RH=50%). As a result of the comprehensive impacts of solar radiation, temperature, relative humidity and wind, the wind tunnel test results showed a much more mutative characteristic and contributed to describe the material in a way that was closer to the practical conditions.

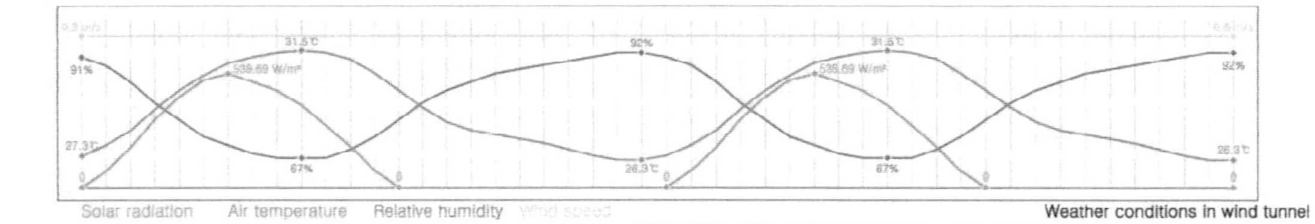

Figure 4. Hourly mass change rate of bamboo scrimber and hardwood specimens of wind tunnel test.

4. Conclusion

1) Nine static test items for hygrothermal properties were carried out for bamboo scrimber. Results showed that, by comparison with reference timbers, bamboo scrimber had higher heat storage and heat transport properties, lower moisture storage and transport properties.
2) A dynamic test in wind tunnel was performed to examine the moisture absorption and desorption rate in practical outdoor climate condition. Results showed that bamboo scrimber had lower moisture absorption and desorption rate than the reference hardwood.
3) The significant magnitude difference between the static and dynamic test results showed the necessity of a comprehensive evaluation approach that could take more practical conditions into consideration. Further works to examine the performance of bamboo scrimber under practical external climate, internal load, construction type, HVAC conditions, etc. were required.

5. Reference

[1] Oscar Hidalgo López 2003 *Bamboo: The Gift of the Gods* (Bogotá, Colombia: O. Hidalgo-Lopez)
[2] Walter Liese and Michael Köhl 2015 *Bamboo: The Plant and Its Use* (New York: Springer)
[3] ZHANG Qisheng, JIANG Shenxue and TANG Yongyu 2002 INBAR *Technical Report No.26: Industrial Utilization on Bamboo* (Beijing: International Network for Bamboo and Rattan)
[4] P.M. Ganapathy, Zhu Huan-Ming, S.S. Zoolagud, D. Turcke and Z.B. Espiloy 1999 INBAR *Technical Report No.12: Bamboo Panel Boards: a State-of-the-Art Review* (Beijing: International Network for Bamboo and Rattan)
[5] LY/T 1660-2006 Standard terminology for bamboo-based panel (in Chinese)
[6] Yimin Sun, Zujian Huang and Wang Pan 2015 Manufacturing Classification of Bamboo and Comparative Study on Several Physical Properties, Specially Considering the Application in Building Envelope *Proc. of the 31st Int. PLEA Conf. Ass* vol 3 (Bologna: Building Green Futures) p 651
[7] Hui LI, Yixin ZHU, Zhibin YANG and Bin HONG 2013 A Review on the Bamboo Microstructure and Bamboo Fiber Application in China *China Forest Science and Technology.* **27(3)** 1-5 (in Chinese)
[8] WU Bingling, YU Yanglun, QI Jinqiu and YU Wenji 2014 Effects of bamboo bundles with fine fluffing and carbonized treatment on the properties of bamboo scrimber *Journal of Nanjing Forestry University Natural Science Edition.* **38(6)** 115-120 (in Chinese)
[9] YU Yanglun, LIU Bo and YU Wenji 2014 Research Progress on New Technology and New Product Development of Bamboo Scrimber *International Wood Industry Special Report.* (in Chinese)
[10] XIAO Zhongping, ZHANG Sujun and SHU Biqing 2013 Application of Carbonized Glued Laminated Bamboo in Bamboo Structure Construction *China Forest Products Industry.* **40(6)** 44-53 (in Chinese)
[11] WANG Chunxia, CUI Lidong, LIU Haoyang, ZHANG Jing and ZHANG Changwu 2013 Study on Cold Press Technology for Reconsolidated Bamboo *Forest Machinery & Woodworking Equipment.* **41(12)** 17-19 (in Chinese)
[12] TANG Liancheng 2012 Study on Manufacturing and Properties of Strand Woven Bamboo (Wood) Lumber Made of Sympodial Species *Dissertation of Nanjing Forestry University* (in Chinese)
[13] AN Xin 2012 Study on Control Techniques of Termite for Sympoial Bamboo and Composite Materials *Dissertation of Chinese Academy of Forestry* (in Chinese)
[14] LI Ren 2014 Preparation and combustion characteristics of frame retardant reconstituted bamboo *Dissertation of Zhejiang A&F University* (in Chinese)
[15] Molenda C.H.A., Crausse P. and Lemarchand D 1992 The influence of capillary hysteresis effects on the humidity and heat coupled transfer in non-saturated porous medium *International Journal of Heat and Mass Transfer.* **35(6)** 1385-96
[16] Pedersen C.R 1992 Prediction of moisture transfer in building constructions. *Building and Environment.* **27(3)** 387-97

[17] Künzel H.M 1995 *Simultaneous Heat and Moisture Transport in Building Components* (Stuttgart: Fraunhofer IRB Verlag Suttgart,
[18] Häupl P., Grunewald J., Fechner H., et al 1997 Coupled heat air and moisture transfer in building structures *International Journal of Heat and Mass Transfer*. **40(7)** 1633-42
[19] Janssen H., Blocken B. and Carmeliet J 2007 Conservative modelling of the moisture and heat transfer in building components under atmospheric excitation *International Journal of Heat and Mass Transfer*. **50(5-6)** 1128-40
[20] Mendes N. and Philippi P. C 2005 A method for predicting heat and moisture transfer through multilayered walls based on temperature and moisture content gradients. *International Journal of Heat and Mass Transfer*. **48(1)** 37-51
[21] Steeman H.J., Van Belleghem M., Janssens A., et al 2009 Coupled simulation of heat and moisture transport in air and porous materials for the assessment of moisture related damage *Building and Environment*. **44(10)** 2176-84
[22] Tariku F., Kumaran K. and Fazio P 2010 Transient model for coupled heat, air and moisture transfer through multilayered porous media *International Journal of Heat and Mass Transfer*. **53(15-16)** 3035-44
[23] Kumaran M.K 1996 *IEA Annex 24: Heat, Air and Moisture Transfer in Insulated Envelope Parts. Final Report, Volume 3, Task 3: Material Properties*
[24] Woloszyn M. and Rode C 2008 *IEA Annex 41: Whole Building Heat, Air, Moisture Response. Subtask 1: Modeling Principles and Common Exercises* (International Energy Agency, Executive committee on Energy Conservation in Buildings and Community Systems)
[25] Van Belleghem M., Steeman H. J., Steeman M., et al 2010 Sensitivity analysis of CFD coupled non-isothermal heat and moisture modelling *Building and Environment*. **45(11)** 2485-96
[26] FENG Chi 2014 Study on the Test Methods for the Hygric Properties of Porous Building Materials *Dissertation of South China University of Technology* (in Chinese)
[27] ASTM D7370-2009: Standard Test Method for Determination of Relative Density and Absorption of Fine, Coarse and Blended Aggregate Using Combined Vacuum Saturation and Rapid Submersion
[28] DIN EN 1936-2007: Natural stone test method - Determination of real density and apparent density, and of total and open porosity
[29] ISO 12571:2012: Hygrothermal performance of building materials and products - Determination of hygroscopic sorption properties
[30] ASTM C1498 - 04a(2016): Standard Test Method for Hygroscopic Sorption Isotherms of Building Materials
[31] ISO 12572:2001(E): Hygrothermal performance of building materials and products - Determination of water vapour transmission properties
[32] ISO 15148:2002(E): Hygrothermal performance of building materials and products - Determination of water absorption coefficient by partial immersion
[33] Scheffler G.A. and Plagge R 2010 A whole range hygric material model: Modelling liquid and vapour transport properties in porous media *International Journal of Heat and Mass Transfer*. **53(1-3)** 286-296
[34] ISO 11357-4-2005: Plastics - Differential scanning calorimetry (DSC) - Part 4: Determination of specific heat capacity
[35] ISO 8302-1991: Thermal insulation; determination of steady-state thermal resistance and related properties; guarded hot plate apparatus
[36] DIN EN 16012-2012: Thermal insulation for buildings - Reflective insulation products - Determination of the declared thermal performance

Acknowledgments
The research is funded by: the (Chinese) State Key Laboratory of Subtropical Building Science Project: 2017KC22; the State Scholarship Fund by the Chinese Scholarship Council: 201506150017.

Assessment on bamboo scrimber as a substitute for timber in building envelope in tropical and humid subtropical climate zones - part 2 performance in building envelope

Zujian HUANG[1,2], Yimin SUN[1] and Florian MUSSO[2]

1 School of architecture, State Key Laboratory of Subtropical Building Science, South China University of Technology, Guangzhou 510640, China
2 Chair of Building Construction and Material Science, Department of Architecture, Technical University of Munich, Munich 80333, Germany
E-mail: arymsun@scut.edu.cn

Abstract. Bamboo scrimber was fiber based panel developed in 2000s that was potential to be an ideal substitute for construction timber in bamboo growing areas. For evaluating the performance of applying bamboo scrimber in building envelope by clarifying the performance difference with timber, softwood (SW) and hardwood (HW) units were set as reference model, accordingly bamboo scrimber (BFB) units of the same construction and space size as evaluation model, by which performance regarding to building component and space unit were compared. Space units enclosed with bamboo, softwood and hardwood exterior walls were constructed and simulated in WUFI Plus for 16 cities from tropical and humid subtropical regions overlapped with the bamboo forest distribution. Bamboo scrimber showed changeable strengths of heat storage and vapor resistance and weakness of heat transport properties, which varied with the external climate, building function, construction type and HVAC condition.

1. Introduction

The world's bamboo forest distribution largely overlapped the tropical and humid subtropical climate zones where existed dense population, large construction timber consumption, meanwhile enduring bamboo tradition and extensive bamboo research accumulation, which made bamboo, in panel or square forms, to be an ideal construction material and potential substitute for timber. Different from the other bamboo based panels, mainly including plybamboo (1980s), bamboo particleboard (1990s), bamboo OSB (1990s) and bamboo laminated lumber (1990s) [1, 2, 3] bamboo scrimber, developed in 2000s, was fiber based panel that could be made from herbaceous and small dimension bamboos, which were widely distributed but hadn't been fully utilized. Thus bamboo scrimber was potential to be promoted in bamboo growing areas to relieve the local timber shortage situation. (Figure 1)
The existing study on the bamboo scrimber from the perspective of building industry mostly focused on the mechanical properties for load bearing application, which showed bamboo scrimber had mechanical properties that were equal to or surpassed timber [4, 5, 6].The construction practices were preceded in China, where scholars and architects took reference from the multilayer timber construction system to carry out the application of bamboo panels in building envelope. Qisheng ZHANG, Zhitao LU developed bamboo anti-seismic housing in 2009, in which various bamboo based panels including bamboo scrimber were used as wall framework with insulation material filled inside to achieve favorable thermal performance [7]. Hao Lin adopted the 2×4 construction method (2×4 inch, 5×10cm) in bamboo housing, in which bamboo laminated lumber and bamboo scrimber were

used as wall framework and insulation material as filler[8]. Various bamboo profiles for flooring, ceiling and wall were applied in building envelope, which formed a parallel and competitive product series with timber in some regions. Vogtlander, J.G. and Van der Lugt, P. carried out LCA assessment on bamboo scrimber flooring products and showed that the long-distance transport would accounted for a major part of the CO_2 footprint and eco-costs [9, Therefore the performance comparison with timber competitive products and adaptability evaluation in local regions were of great significance.

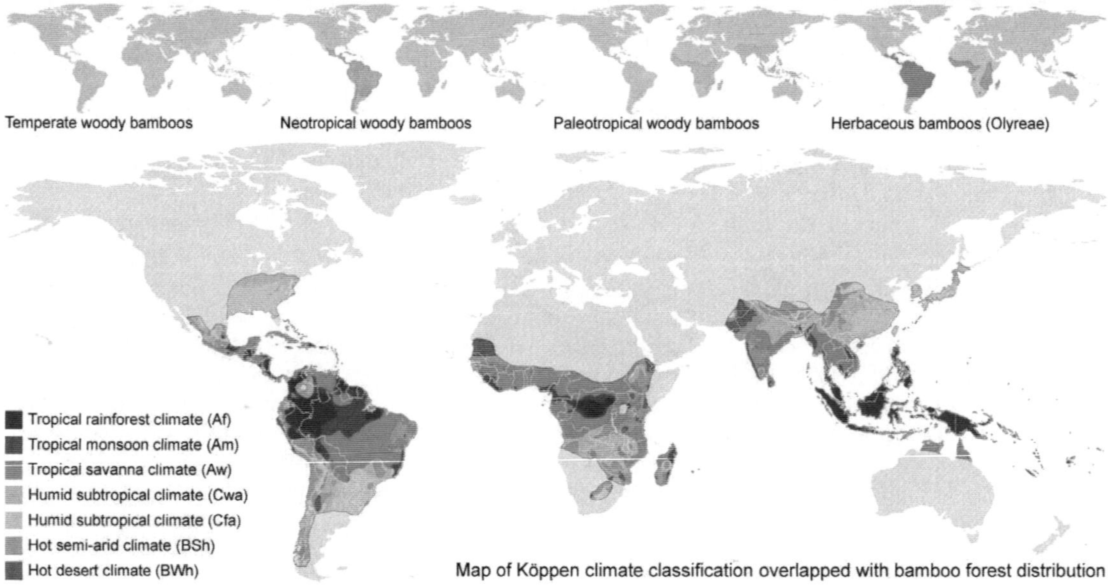

Figure 1. Bamboo forest distribution and the overlapping with the climate zones.

Based on the hygrothermal properties test, bamboo scrimber, softwood and hardwood were designed for exterior walls with same construction size and compared for thermal and hygric performance, afterwards enclosed space units with the above exterior walls were constructed in WUFI Plus and simulated the annual dynamic hygrothermal performance of the exterior walls, indoor environment and HVAC demand. The test, calculation and simulation results could provide support for practical application, with consideration to climate condition, building function, construction type, etc.

2. Simulation model design

The simulation models were constructed in WUFI Plus, which were composed of external conditions, internal conditions, boundary conditions and HVAC conditions: (Figure 2)

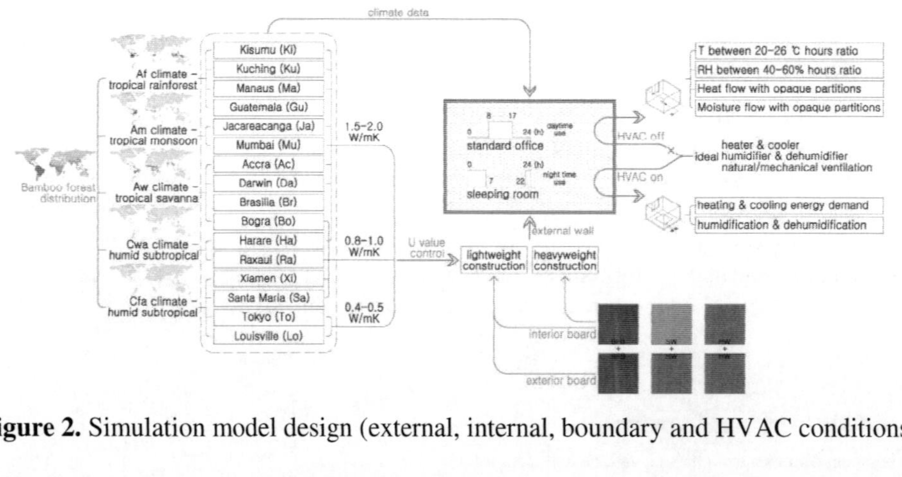

Figure 2. Simulation model design (external, internal, boundary and HVAC conditions).

IOP Conf. Series: Materials Science and Engineering **264** (2017) 012007 doi:10.1088/1757-899X/264/1/012007

2.1. External conditions

According to Köppen-Geiger climate classification, 16 cities in 5 groups were chosen as the representative cities for Tropical rainforest climate zone (Af), Tropical monsoon climate zone (Am), Tropical savanna climate zone(Aw), Humid subtropical climate zone (Cwa and Cfa). The TMY weather data of the 16 cities were input as the external conditions for the model. (Table 1)

Table 1. Distribution of the 16 representative cities.

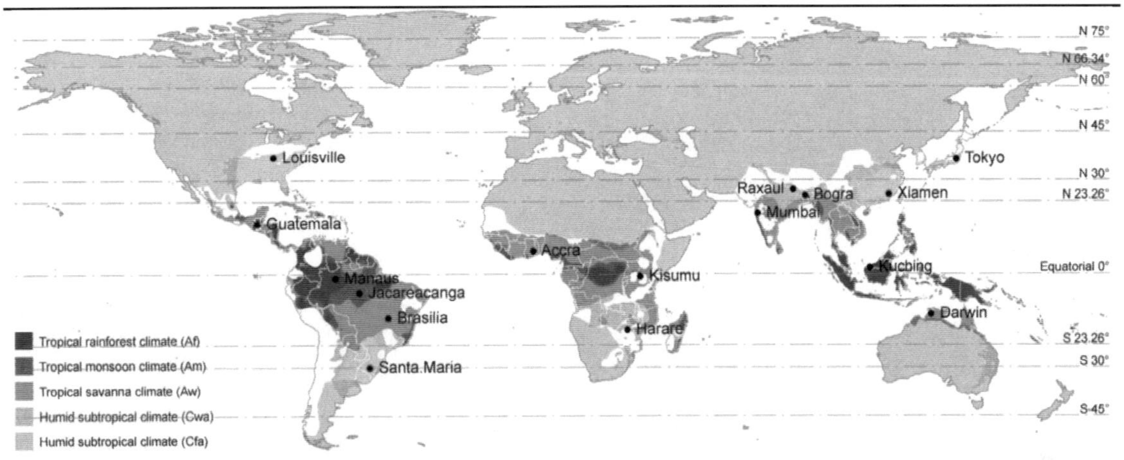

Climate zone	Sub zone	City	Latitude	Longitude	Continent
Tropical climate	Af - Tropical rainforest climate	Kisumu	-0.091702	34.767957	Africa
		Kuching	1.607681	110.37854	Asia
		Manaus	-3.119028	-60.021731	South America
		Guatemala	15.783927	-90.23037	Central America
	Am - Tropical monsoon climate	Jacareacanga	-7.316864	-57.462726	South America
		Mumbai	19.075984	72.877656	Asia
	Aw - Tropical savanna climate	Accra	5.603717	-0.186964	Africa
		Darwin	-12.46344	130.84564	Australia
		Brasilia	-15.794157	-47.882529	South America
		Bogra	24.843559	89.370108	Asia
Humid subtropical climate	Cwa - Humid subtropical climate	Harare	-17.825166	31.03351	Africa
		Raxaul	26.979768	84.85158	Asia
	Cfa - Humid subtropical climate	Xiamen	24.479833	118.08943	Asia
		Santa.Maria	-29.686511	-53.817054	South America
		Tokyo	35.689487	139.69171	Asia
		Louisville	38.252665	-85.758456	North America

2.2. Internal conditions

Two types of building functions were set. Type O was office building represented by standard office that was 3.0m×6.0m×3.0m (width×depth×height). The occupancy schedule was 8:00 - 17:00, which represented the space for daytime use. Type R was residential building represented by sleeping room that was 3.0m×4.0m×3.0m (width×depth×height). The occupancy schedule was 22:00 - 07:00, which represented the space for night time use.

2.3. Boundary conditions, namely the building envelope

The exterior walls were chosen as the study objects, while the flooring and ceiling were set as partitions between the same interior conditions. Considering that in practical projects, the bamboo/timber boards were generally used in lightweight or heavyweight constructions, both types with approximate U values were designed. (Table 2)

For ensuring the comparability among different variables, the exterior walls were designed into 3 groups of gradient U values by adjusting single factors. From bamboo to timber components: replaced the bamboo scrimber (BFB) with softwood (SW) and hardwood (HW) boards, while kept the remaining construction layers; From lightweight construction to heavyweight construction types: replaced the 18mm exterior board and the insulation layer with dena Brick 800 of the same thermal resistance value, while kept the remaining construction layers; The 3 groups of components were assigned to the 16 cities mainly according to the climate zones and the corresponding latitudes.

Table 2. Boundary condition (Construction of the external walls).

Group	Type	Exterior						Interior			Construction graph	
		1-1		1-2	3	4	5	6				
		BFB	HW	dena Brick 800	PE foil	EPS board	Air	BFB	SW	HW	n- light construction	r-heavy construction
U 1	BFB-n	18.0			0.002	0	40.0	12.5			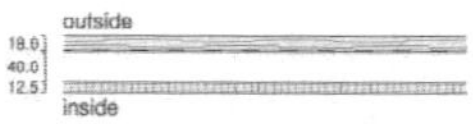	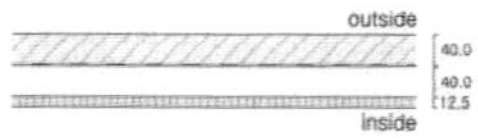
	SW-n		18.0		0.002	0	40.0		12.5			
	HW-n		18.0		0.002	0	40.0			12.5		
	BFB-r			40.0	0.002		40.0	12.5				
	SW-r			40.0	0.002		40.0		12.5			
	HW-r			40.0	0.002		40.0			12.5		
U 2	BFB-n	18.0			0.002	20.0	40.0	12.5			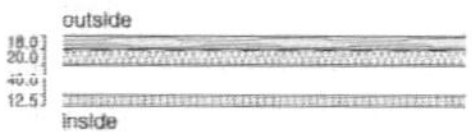	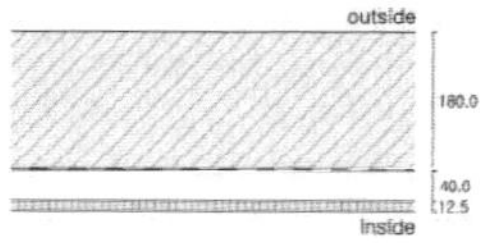
	SW-n		18.0		0.002	20.0	40.0		12.5			
	HW-n		18.0		0.002	20.0	40.0			12.5		
	BFB-r			180.0	0.002		40.0	12.5				
	SW-r			180.0	0.002		40.0		12.5			
	HW-r			180.0	0.002		40.0			12.5		
U 3	BFB-n	18.0			0.002	60.0	40.0	12.5			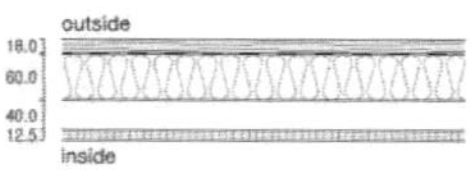	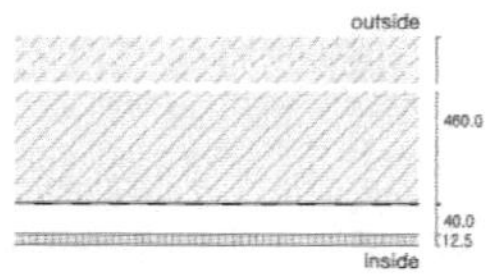
	SW-n		18.0		0.002	60.0	40.0		12.5			
	HW-n		18.0		0.002	60.0	40.0			12.5		
	BFB-r			460.0	0.002		40.0	12.5				
	SW-r			460.0	0.002		40.0		12.5			
	HW-r			460.0	0.002		40.0			12.5		

n-lightweight construction, r-heavyweight construction;
The group1, 2 and 3 were designed with U value control of 1.5-2.0W/mK, 0.8-1.0W/mK and 0.4-0.5W/mK

2.4. HVAC conditions

Based on the models consisted of the above conditions, the HVAC were firstly turned off to evaluate hygrothermal performance of the exterior walls and the indoor environment, which was characterized by the indicators including heat and moisture flow, interior surface and air temperature, relative humidity, etc. during the occupancy hours. Afterwards the HVAC were turned on and ran the model with ideal heater & cooler, humidifier & dehumidifier and mechanical ventilation to maintain certain thermal comfort, and calculated the annual heating & cooling, and the humidification & dehumidification demand.

3. Results analysis

3.1. Building component static calculation results

Six indicators related to the thermal and hygric performances were chosen and compared between bamboo and timber building components. The heat transfer coefficient - U value was the most commonly used indicator to characterize the thermal transport performance. The heat capacity - S value, and thermal capacity inside - S_i value were indicators to characterize the thermal stability, which had decisive effect on the temperature phase shift, attenuation and fluctuation on interior surface. The vapor diffusion thickness - s_d value was chosen to characterize the total vapor permeation resistance of the whole components. (Table 3, Figure 3)

3.1.1. Heat transport properties. Calculated results showed that with the same construction dimensions and layer arrangement, the U values of bamboo components were higher than the timber components, which was caused by the higher thermal conductivity of bamboo scrimber. From U1-U3, the Bamboo-Timber U value magnitudes were narrowed from 0.058-0.270 W/m^2K to 0.004-0.021 W/m^2K.

3.1.2. Heat storage properties. Calculated results showed that the S values and S_i values of bamboo components was higher than the timber components, with magnitudes correspondingly of 10-30 KJ/m^2K and 6.1-13.4 KJ/m^2K. The magnitudes were larger between bamboo scrimber and softwood components than that between bamboo scrimber and hardwood components, and meanwhile larger in lightweight construction than in heavyweight construction. The temperature phase shifts of bamboo components were 0.7-2.0 h longer than timber components. The temperature attenuation magnitudes of bamboo components to timber components of Group U1 and U2 were 0.082-0.874 slightly larger than timber components, but were enlarged significantly to 1.194-1.652 in lightweight construction and 7.937-12.078 in heavyweight construction in U3.

3.1.3. Hygric performance. Calculated results showed that the s_d values of bamboo scrimber components was larger than the timber components, with magnitudes of 11.62 m in lightweight construction and 4.77 m in heavyweight construction, which was caused by the higher water vapor permeation resistance factor - μ value of bamboo scrimber. It meant that the bamboo components were of higher resistance to moisture transport, which could be dialectical that could help to weaken the vapor transport on one side, while on the other side might slow down the indoor moisture from exhausting.

Table 3. Building components thermal and hygric performance calculation results.

Group	Type	Heat transfer coefficient	Heat capacity	Thermal capacity inside	Phase shift	Temperature attenuation	Vapor diffusion thickness
		[W/m^2K]	[KJ/m^2K]	[KJ/m^2K]	[h]	[-]	[m]
U1	BFB-n	1.880	53	17.1	2.3	1.130	23.64
	SW-n	1.610	23	6.7	1.1	1.022	12.02
	HW-n	1.727	28	9.5	1.5	1.048	12.02
	BFB-r	1.773	48	17.2	2.5	1.159	16.22
	SW-r	1.597	33	8.3	1.0	1.030	11.45
	HW-r	1.715	37	11.1	1.7	1.057	11.45
U 2	BFB-n	0.969	53	19.3	4.0	1.637	24.24
	SW-n	0.892	24	7.1	2.0	1.101	12.62
	HW-n	0.927	28	10.5	2.7	1.233	12.62
	BFB-r	0.940	139	55.0	8.0	3.731	18.32
	SW-r	0.888	124	42.0	7.0	2.857	13.55
	HW-r	0.923	129	47.0	7.3	3.096	13.55
U 3	BFB-n	0.492	54	21.0	5.3	3.077	25.44
	SW-n	0.471	25	7.6	3.5	1.425	13.82
	HW-n	0.481	29	11.5	4.3	1.883	13.82
	BFB-r	0.484	321	146.0	17.5	55.556	22.52
	SW-r	0.470	306	130.0	16.5	43.478	17.75
	HW-r	0.480	311	136.0	16.8	47.619	17.75

n-lightweight construction, r-heavyweight construction

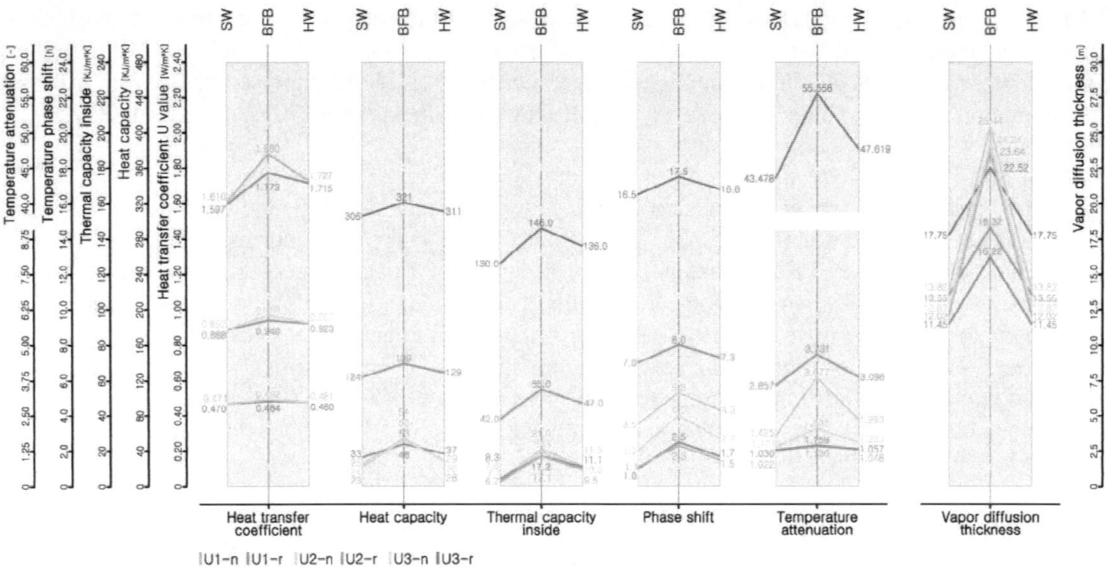

Figure 3. Thermal and hygric performance comparison between bamboo and timber components.

3.2. Enclosed space units dynamic simulation results

The enclosed space units were simulated for the whole year in both HVAC on and off condition. The hygrothermal behavior of the exterior walls and the indoor environment during the occupancy hours without HVAC, afterwards the annual heating & cooling, humidification & dehumidification demand with ideal HVAC were analyzed. The BFB-SW and BFB-HW ratios were calculated to show the strengths and weakness and evaluate how far the difference between bamboo and timber were.

3.2.1. Hygrothermal performance of the exterior walls. In HVAC off condition, the heat exchange with exterior walls and the interior surface temperature amplitude were chosen as indicators to characterize the thermal performance. As for the heat exchange, due to the higher U value, the BFB units were mostly up to 46.65% and 13.30% larger than SW and HW units, but in certain cases, mostly the residential space, the heat exchange with BFB units was lesser. The interior surface temperature of BFB components was up to 12.62% and 6.35% lower than the SW and HW units, which resulted from the higher S value. Moisture exchange with the exterior walls and the total moisture flow were chosen as indicators to characterize the hygric performance (The former value was remainder of the moisture gain and moisture loss, while the latter was the sum of all the moisture flow through the components). Due to the higher μ value, the total moisture flow through BFB components were up to 33.54% smaller than the SW and HW units, which was beneficial for preventing the building component from moisture damage. However, the moisture exchange, which was affected by both the μ value and the vapor transfer direction, didn't show that clear regularity. Mostly the moisture exchange of BFB components were 29.26% and 43.15% lesser that SW and HW units, while in some cases, mostly the SW residential space, the values of BFB components were 48.81% and 32.23% larger than the SW and HW units. (Figure 4)

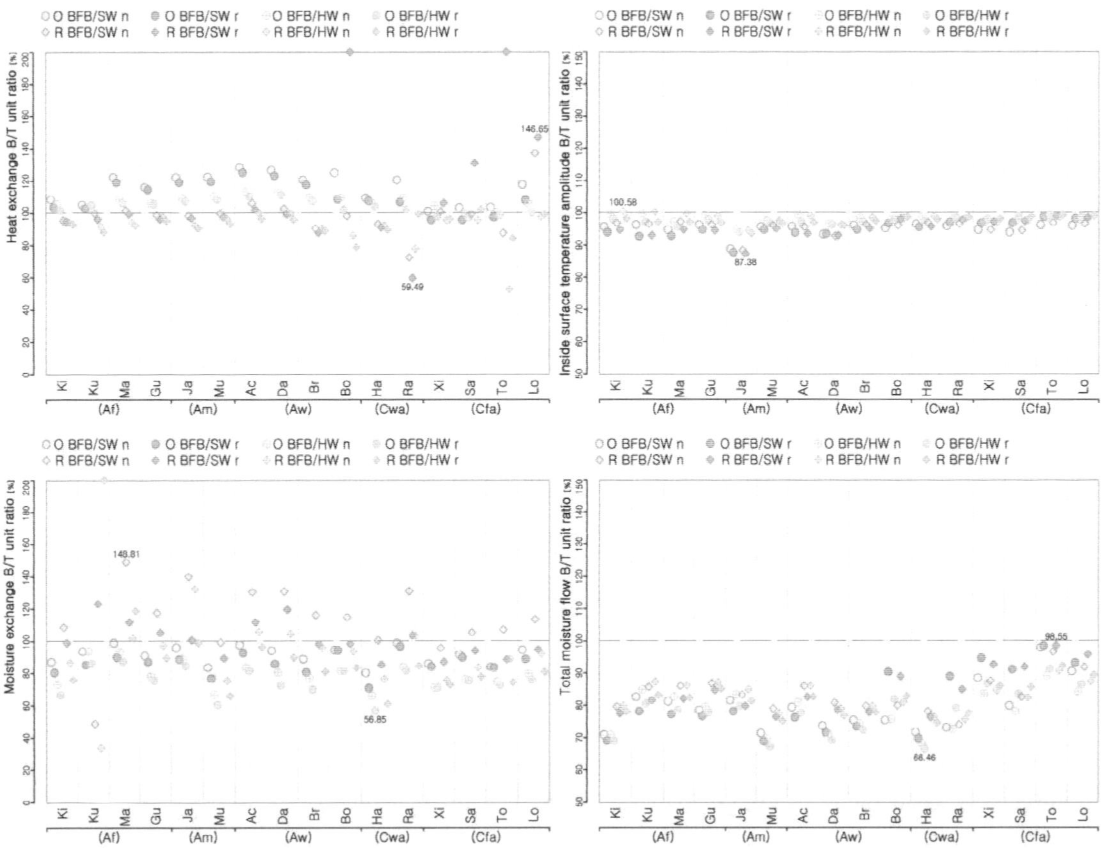

Figure 4. Exterior walls hygrothermal performance B/T ratio (due to small divisor in some cases, few B/T values might be particularly large, which were presented on the upper boundaries of the graph and weren't counted as extreme values).

3.2.2. Indoor environment. In HVAC off condition, the hours proportion in occupancy period of the interior air temperature between 20-26℃, and the interior relative humidity between 40%-60% were chosen as indicators to describe the hygrothermal environment of the enclosed space units. As for the interior temperature, the BFB units performed better than SW and HW units, corresponding with magnitudes up to 112.68% and 63.16%, but in certain cases, mostly the residential space, the BFB units performed worse than SW and HW units, corresponding with magnitudes up to 20.81% and 10.57%. The difference magnitudes between BFB and SW were larger than between BFB and HW, and in lightweight construction larger than in heavyweight construction, and both narrowed with thermal performance improvement of the whole components. As for the relative humidity, due to the continuous high relative humidity, in most cases there were very few or even none hours when the indoor relative humidity could maintain within 40%-60%, which led to incomparability between BFB and timber units. Probably, the BFB units performed worse in Cwa zones and better in Cfa zones than SW and HW units, and approximately in Af, Am and Aw zones. (Figure 5)

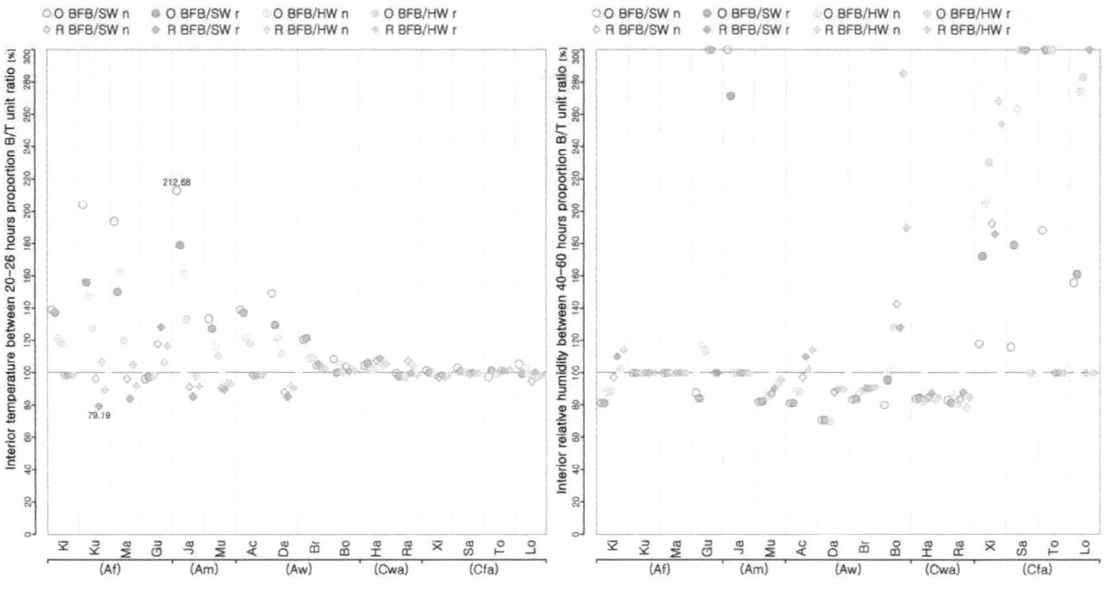

Figure 5. Indoor environment performance B/T ratio (due to small divisor in some cases, few B/T values might be particularly large, which were presented on the upper boundaries of the graph and weren't counted as extreme values, when all the BFB, SW and HW values were 0, the B/T ratio were regarded as 100%).

3.2.3. HVAC demand. In HVAC on condition, ideal heater & cooler, humidifier & dehumidifier and mechanical ventilation were arranged to maintain certain indoor comfort. The annual total energy, humidification & dehumidification demand were indicators for comparison. The energy demand of BFB units was mostly up to 22.22% larger than SW units, and 16.87% larger than HW units due to the higher U values of BFB components, however in Kisumu and Brasilia, the BFB units could even consume lesser energy, which might result from the raising impact of S values. In aspect of hygric performance, the BFB units required mostly up to 12.84% smaller humidification & dehumidification demand than SW and HW units, while in particular cases the BFB units would require up to 24.17% larger than SW and HW units, which meant that the higher s_d value of BFB components didn't ensure smaller humidification & dehumidification demand that was also affected by the vapor transfer direction. In hot and humid regions the higher s_d value of BFB components could help to buffer the influence of outside humid air on the indoor environment, while in some cases it might also block the indoor moisture from exhausting. (Figure 6)

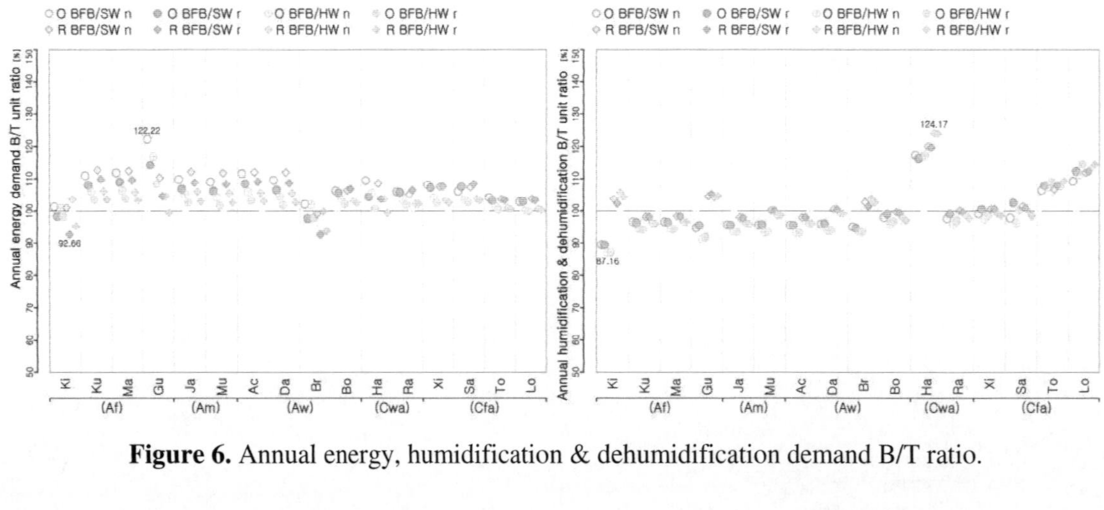

Figure 6. Annual energy, humidification & dehumidification demand B/T ratio.

IOP Conf. Series: Materials Science and Engineering **264** (2017) 012007 doi:10.1088/1757-899X/264/1/012007

4. Conclusion

1) Static test and calculation results showed that, compared with the timber units, bamboo scrimber units had advantages in heat storage performance that provided stable indoor thermal environment particularly in non HVAC condition and lightweight construction; stronger vapor resistance that was beneficial for preventing from damage caused by moisture flow; on the other hand, the higher heat transport performance of bamboo scrimber components should be improved.
2) The dynamic annual simulation in WUFI Plus showed that thermal and hygric performance above had conclusive impact on hygrothermal performance of the enclosed space, however the weights of each indicators varied with climate condition, building function, construction type and HVAC control.
3) The difference between the static calculation and dynamic simulation results showed the necessity of a comprehensive evaluation approach that could cover progressive levels on the physical performance of material, building component and enclosed space. Verification for the simulation results by either test houses or practical application feedbacks were required in further works.

5. Reference

[1] P.M. Ganapathy, Zhu Huan-Ming, S.S. Zoolagud, D. Turcke and Z.B. Espiloy 1999 *INBAR Technical Report No.12: Bamboo Panel Boards: a State-of-the-Art Review* (Beijing: International Network for Bamboo and Rattan)
[2] LY/T 1660-2006 Standard terminology for bamboo-based panel (in Chinese)
[3] Yimin Sun, Zujian Huang and Wang Pan 2015 Manufacturing Classification of Bamboo and Comparative Study on Several Physical Properties, Specially Considering the Application in Building Envelope *Proc. of the 31st Int. PLEA Conf. Ass* vol 3 (Bologna: Building Green Futures) p 651
[4] Bhavna Sharma, Ana Gatóo, Maximilian Bock and Michael Ramage 2015 Engineered bamboo for structural applications *Construction and Building Material.* **81** 66-73
[5] Anuj Kumar, Tomáš Vlach, Lenka Laiblova, Martin Hrouda, Bohumil Kasal, Jan Tywoniak and Petr Hajek 2016 Engineered bamboo scrimber: Influence of density on the mechanical and water absorption properties *Construction and Building Material.* **127** 815-827
[6] Zhou Aiping, Huang Dongsheng, Li Haitao and Su Yi 2012 Hybrid approach to determine the mechanical parameters of fibers and matrixes of bamboo *Construction and Building Material.* **35** 191-196
[7] Wei Yang, LU Qingfang, Zhang Qisheng, Yu Yongzhe and LU Zhitao 2009 Design and Construction of Modern Bamboo Anti-seismic Living Room *Construction Technology.* **38(11)** 52-54 (in Chinese)
[8] HAO Lin 2010 Contemporary Bamboo Construction for A Green Future *Eco-city and Green Building.* **4** 80-89 (in Chinese)
[9] P van der Lugt, Joost Vogtländer and Han Brezet 2008 Bamboo, a Sustainable Solution for Western Europe - Design Cases, LCAs and Land-use *INBAR Technical Report No. 30* (Beijing: International Network for Bamboo and Rattan)
[10] P Van der Lugt, A Van den Dobbelsteen and JJA Janssen 2006 An environmental, economic and practical assessment of bamboo as a building material for supporting structures *Construction and Building Materials.* **20 (9)** 648-656

IOP Conf. Series: Materials Science and Engineering **264** (2017) 012008 doi:10.1088/1757-899X/264/1/012008

Properties of concrete containing ground palm oil fuel ash as fine aggregate replacement

W A. Saffuan[1], K Muthusamy[2], N. A. Mohd Salleh[3] and N. Nordin[4]

1,2,3 Faculty of Civil Engineering and Earth Resources Universiti Malaysia Pahang, Lebuhraya Tun Razak, 26300, Gambang, Pahang
4 Faculty of Engineering and Technology Infrastructure, Department of Civil Engineering, Infrastructure Universiti Kuala Lumpur, 43000 Kajang, Selangor
khairunisa@ump.edu.my

Abstract. Environmental degradation resulting from increasing sand mining activities and disposal of palm oil fuel ash (POFA), a solid waste generated from palm oil mill needs to be resolved. Thus, the present research investigates the effect of ground palm oil fuel ash as partial fine aggregate replacement on workability, compressive and flexural strength of concrete. Five mixtures of concrete containing POFA as partial sand replacement designed with 0%, 10%, 20%, 30% and 40% of POFA by the weight of sand were used in this experimental work. The cube and beam specimens were casted and water cured up to 28 days before subjected to compressive strength and flexural strength testing respectively. Finding shows that concrete workability reduces as the amount of POFA added become larger. It is worth to note that 10% of POFA is the best amount to be used as partial fine aggregate replacement to produce concrete with enhanced strength.

1. Introduction

Currently, in modern industrial society, the concrete is expansively used with more than 10 billion tons produced every year [1] Natural sand is prime material used for the preparation of concrete. With the growth in urbanization and industrialization, the demand for sand supply is increased day by-days. Continuous sand mining from the natural environment will cause the depletion of this material in the future generation. The reduction in the sources of natural sand and the requirement for lowering the cost of concrete production tend to find new alternative material to replace natural sand in concrete. In order to resolve these problems, one of the available materials that can be widely found in Malaysia is palm oil fuel ash (POFA). Utilization palm oil fuel ash (POFA) as partial fine aggregate replacement in concrete is seen as one of the steps to reduce the use of natural sand in concrete production.

Palm oil fuel ash (POFA) is an agro-waste ash results from palm oil residues, such as palm fiber and shells. These residues being burned at temperatures of about 800°C -1000°C to produce steam for electricity generation in biomass thermal power plants [2]. Malaysia produced about 3 million tons of POFA in 2007 [3]. This production rate is likely to increase due to the increasing plantation of palm oil trees [4]. In practice, this solid waste is dumped near the palm oil mill inside the factory area. This light greyish ash is easy to be carried away by the wind thus creating environmental pollution. Past researcher [5] highlighted that the negative impact of POFA dumping to the environment is evident. Thus, approach of integrating this waste material in concrete production would be able to reduce amount of waste ending at landfill and consumption of natural river sand.

2. Experimental programme

2.1. Materials

The materials used were ordinary Portland cement Type I (OPC), river sand, coarse aggregate, water and Palm Oil Fuel Ash (POFA). The river sand were supplied by local manufacturer nearby. Coarse aggregate used is crushed granite aggregate. Supplied tap water at the laboratory were used for concrete mixing and curing purpose. Palm Oil Fuel Ash (POFA) was supplied from palm oil mill factory located in Ladang Lepar Baru, West Malaysia. The POFA at dumping area as shown in Figure 1 was collected using shovel and then packed tightly in gunny. After that, it is brought to the laboratory for further processing. At the laboratory, the raw Palm Oil Fuel Ash (POFA) was sieved and then ground to obtain finer POFA before it is used as partial sand replacement material. The processed samples of palm oil fuel ash are illustrated in Figure 2.

Figure 1. POFA dumped at palm oil mill.

Figure 2. Ground palm oil fuel ash.

2.2. Mix proportions

The mixes are plain concrete (control specimen) and four mixes concrete containing various POFA content as fine aggregate replacement. The mixing ingredients of control specimen consists OPC, 100% sand, coarse aggregate and water. While the mixing ingredients of POFA concrete are almost same as plain concrete but the sand has been partly replaced with 10%, 20%, 30% and 40% of ground POFA content by weight. The mix proportion of mixes used is tabulated in Table 1.

Table 1. Mix proportion of concrete mixes in kg/ m^3.

Water (kg)	OPC (kg)	C/W	Sand (kg)	Granite (kg)	POFA (kg)
245	350	0.7	980	850	0
245	350	0.7	882	850	98
245	350	0.7	784	850	196
245	350	0.7	686	850	294
245	350	0.7	588	850	392

2.3. Specimen preparation and testing

All the mixes of fresh concrete were casted into standard cubes sized (100 x 100 x 100 mm) and beam specimen size (100 x 100 x 500 mm) then removed from the mould after 24 hours. All specimens were subjected to water curing as illustrated in Figure 3 for 7, 14 and 28 days. Slump test were conducted to determine the concrete mixes workability by following the procedure stated in BS EN 12350 [6]. All testing procedures performed according to BS EN 12390-3:2002 [7] and ASTM 12390 – 5 [8] for compressive strength and flexural strength respectively.

Figure 3. Specimens subjected to water curing.

3. Result and discussion

3.1. Slump test

The result obtained from figure 4 shows that the amount of ground POFA used as partial sand replacement in concrete influences the workability of fresh concrete. The workability of concrete containing palm oil fuel ash (POFA) reduces as the percentage of POFA used in the mix become larger. The slump value for control specimens was the highest and observed to be medium degree of workability. The concrete slump value at 20% and 30% of POFA replacement had given lower workability with 32mm and 29mm respectively as compared to control specimen which is 50mm. The lowest slump value was recorded when 40% POFA is added as partial sand replacement is 27mm. The reduction in concrete workability when larger quantity of POFA is used is likely due to the physical characteristic of POFA which is finer in size compared to sand. The influence of aggregate's physical characteristic on the workability of concrete has been highlighted by past researcher [9]. Finer particle possess larger surface area thus requiring larger amount of water to coat its surface. Another possible explanation is the higher porosity POFA particles absorbing more water as compared to sand. Past researcher [10], noted the presence of pores in unground POFA in contrast to to solid dense river sand causes reduction in the concrete workability. Study elsewhere which investigates the workability of concrete mix containing POFA as partial cement replacement also reported the similar trend of result.

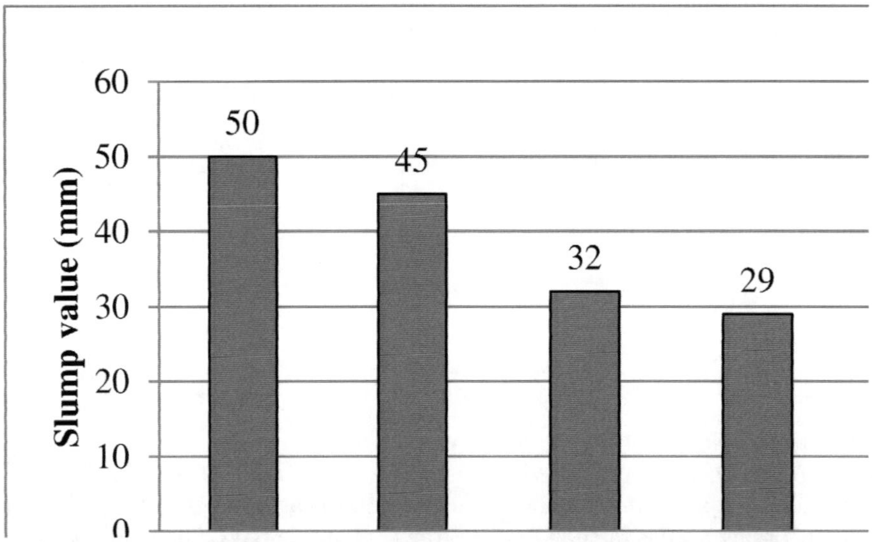

Figure 4. The effect of POFA content on workability of concrete.

3.2. Compressive strength

Figure 5 show the result of compressive strength test of concrete containing POFA. Inclusion of suitable percentage of POFA contributes towards strength enhancement of concrete. All mixes continue to exhibit strength increment as curing age become longer. However, the highest compressive strength was obtained by mix produced using 10% of POFA as sand replacement that exhibit almost 11.31 % higher than control specimens. The increase in compressive strength of 10% palm oil fuel ash (POFA) concrete could possibly due to some factors which is high silica content and high fineness. The chemical reaction between silica from the ash with calcium hydroxide have produced more secondary calcium silicate hydrate gels (C-S-H) in strength development thus improved the concrete strength. According to [11], pozzolanic materials contain high percentage of silica produced more C-S-H gels and reduced the amount of calcium hydroxide that making the concrete stronger, denser and more durable. In addition, high fineness of POFA particle provides a filler effect by filling in the existing voids between sand and aggregate which make the concrete internal structure more packed and contribute to the increment of compressive strength without any chemical reaction [4].

However, a decrease in compressive strength of concrete was observed as the percentage of POFA replacement with sand at 20%, 30% and 40%. In most mixtures, the production of concrete with 40% POFA replacement performed the lowest value compared to control specimens with 14.77MPa. This is because a deficiency in sand will cause the mix of concrete become harsh and contributes to segregation. Increasing in POFA content would increases it fineness. High fineness of POFA particles in the mix will require high water demand due to large surface area of the ash particles. Thus, at constant water-cement ratio, the water are not enough for complete hydration of the cement. And as a result, the quantity of calcium hydroxide $Ca(OH)_2$ will be less, and as a consequence, less secondary calcium silicate hydrate (C-S-H) gel would be produced which plays an important role in making more dense concrete when reacting with the silica contained in the POFA.

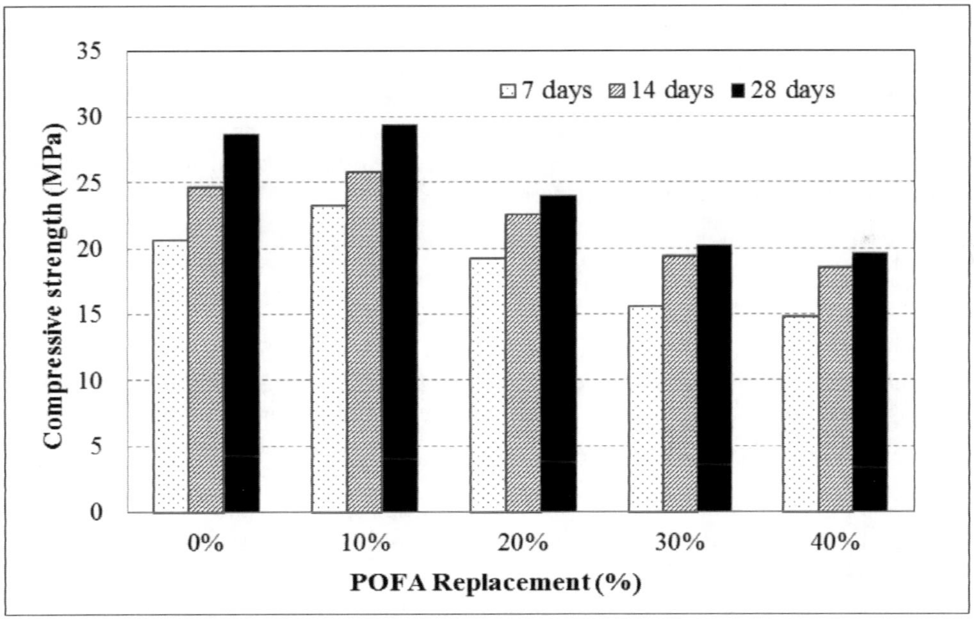

Figure 5. Compressive strengths variation with different percentages of POFA.

3.3. Flexural strength

Figure 6 presents the flexural strength of concrete mixtures containing 0%, 10%, 20%, 30% and 40% of POFA as partial sand replacement. The flexural strength result shows similar trend of the compressive strength. Flexural strength The highest flexural strength was obtained at 10% of sand replacement with POFA that appears to almost 21% higher than control specimens. However, a decreasin trend of flexural strength of concrete can be seen when POFA replacement more than 10%.

The production of concrete with 40% POFA replacement performed the lowest value compared to control specimens with 5.78MPa.

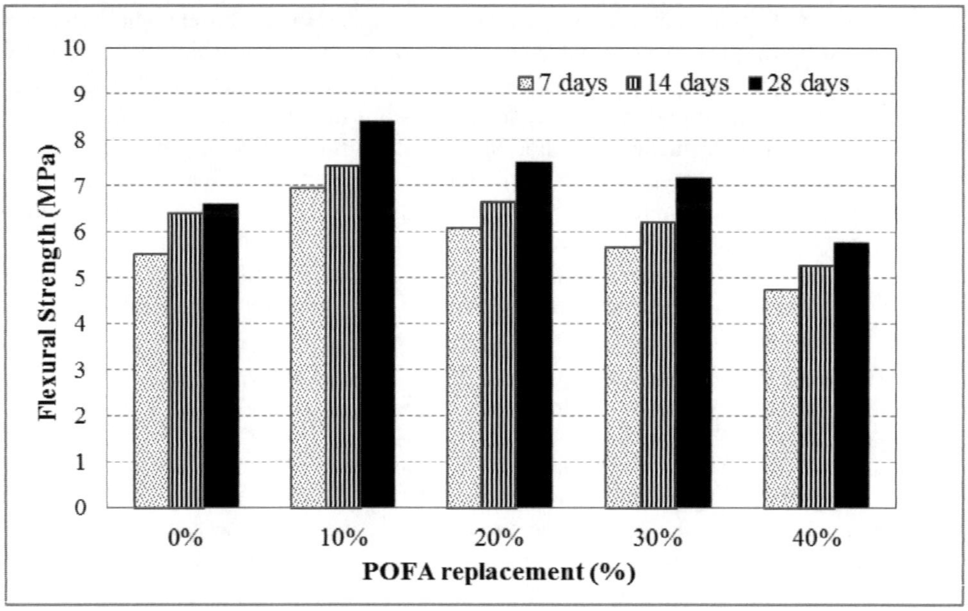

Figure 6. Flexural strength of concrete containing different percentages of POFA.

4. Conclusions

Based on the results obtained from laboratory investigation, it can be concluded that inclusion of suitable percentage of ground POFA as fine aggregate replacement successfully enhances both the compressive strength and flexural strength performance. This finding encourages the use of palm oil waste materials as fine aggregate replacement in concrete, which saves the use of natural sand mined from the river. In addition, amount of palm oil fuel ash disposed at landfill would be reduced thus, assisting the palm oil industry to be more environmental friendly.

5. References

[1] Meyer, C. The greening of the concrete industry. 2009. *Cement and Concrete Composites*, 31, 8, 601-605

[2] Alengaram, J U, Al Muhit, B A and Zamin, M J , 2013. Utilization of oil palm kernel shell as lightweight aggregate in concrete-a review. *Construction. Building. Material*. 38, 161-172

[3] Abdul Awal A S M and Hussin M W. 1997. The effectiveness of palm oil fuel ash in preventing expansion due to alkali–silica reaction. *Cement and Concrete Composites.* 19:367–72

[4] Chindaprasirt P, Homwuttiwong, S and Jaturapittakul C. 2007. Strength and water permeability of concrete containing palm oil fuel ash and rice husk-bark ash. *Construction. Building. Material*. 21:1492–9

[5] Hussin, M W, Muthusamy, K and Zakaria, F. 2010. Effect of mixing constituent towards engineering properties of POFA cement -based aerated concrete. *Journal of Materials in Civil Engineering,* 287-295

[6] BS EN12350 Testing fresh concrete, Compressive strength of test specimens. London. BS EN 12350

[7] BS EN 12390-3 (2009). Testing hardened concrete, Compressive strength of test specimens. London. BS EN 12390-3

[8] BS EN 12390-5 (2009). Testing hardened concrete, Flexural Strength of Test Specimens, London, BS EN 12390-5

IOP Conf. Series: Materials Science and Engineering **264** (2017) 012008 doi:10.1088/1757-899X/264/1/012008

[9] Neville, A M (2011). Properties of Concrete (5th Edition).Prentice Hall

[10] Muthusamy, K, and Zamri, N A. 2014. Exploratory study of palm oil fuel ash as partial cement replacement in oil palm shell lightweight aggregate concrete, *Res. J. Appl. Sci. Eng. Technol.* 8(2) 150-152

[11] Jitchaiyaphum. Sinsiri and Jaturapitakkul, 2013. Cellular lightweight concrete containing high-calcium fly ash and natural zeolite. *International Journal of Minerals, Metallurgy and Materials.* 20, 462-471

IOP Conf. Series: Materials Science and Engineering **264** (2017) 012009 doi:10.1088/1757-899X/264/1/012009

Effects of material properties of HFDFRCC Using recycled fine aggregate on shear strength of RC beam

Yuhei Shiratori[1] and Ken Watanabe[2]

1 Tokai University/Department of Architecture, Hiratsuka City, Japan
E-mail: yuhei1228kenntiku@gmai.com
2 Tokai University/Department of Architecture, Hiratsuka City, Japan
E-mail: kenw@keyaki.cc.u-tokai.ac.jp

Abstract. In this study, we performed loading tests on RC beam specimens made of high-fluidity ductile-fiber-reinforced cementitious composites incorporating recycled fine aggregate with different water-binder ratios. We also performed nonlinear finite element analyses to investigate the effects of water-binder ratios and shear reinforcement bars on RC beam shear strength. Additionally, for some factors, we investigated the influence of the presence or absence of shear reinforcement bars on the shear strength of R-HFDFRCC RC beams. We found that fluctuations in the maximum load of the RC beam specimens attributable to differences in the water-binder ratio can generally be predicted if we understand the differences in the material properties (mainly compressive strength, tensile strength and ultimate tensile strain) of the tough, highly fluid cement compound materials incorporating recycled fine aggregate.

1. Introduction

In recent years, ductile-fiber-reinforced cementitious composites ("DFRCC" hereinafter) with a much better performance than existing fiber-reinforced concrete have been developed [1]. DFRCC are composites of cementitious material reinforced with fibers, which display multiple cracking characteristics and significantly improved toughness when subjected to bending, tension, and compression fracture [1]. These materials overcome the problem of the brittleness of ordinary concrete and are expected not only to significantly improve the performance and durability of concrete structural elements, but to lead to new applications: examples include use in place of conventional cementitious materials as high performance repair materials, shock absorbers, and the like. However, while actual construction examples using DFRCC have been reported [2], they are not common. The reasons include workability issues and high cost compared to other materials. Promoting the use of DFRCC in the future will require the development of new materials, including making improvements to existing materials.

Protecting the global environment during production activities poses a major challenge. In the field of concrete, significant research is being done on recycled aggregate concrete made using recycled aggregate extracted from dismantled concrete blocks. Promoting the expanded use of concrete recycling will require the development of new technologies that allow more effective use of recycled aggregate. Against this backdrop, one of the authors [3], with the goal in mind of improving the workability of DFRCC and expanding the application of recycled fine aggregate, examined the material properties of high fluidity DFRCC ("HFDFRCC" hereinafter) using fly ash as an admixture and fine aggregate as recycled fine aggregate. The results showed that the HFDFRCC incorporating recycled fine aggregate ("R-HFDFRCC" hereinafter) offers sufficient crack dispersion capabilities and excellent bending toughness.

In applying this kind of material in reinforced concrete ("RC" hereinafter) structures, we must identify a shear strength calculation formula capable of appropriately assessing the shear strength of RC members made from R-HFDFRCC. This, in turn, requires performing loading tests of RC members made from R-HFDFRCC. (Compressive strength, tensile reinforcement ratio, shear reinforcement ratio, and shear span ratio are regarded to affect RC member shear strength as test items.) It also requires an accumulation of experimental data based on appropriate strength controls. At the same time, applying analytical techniques, we must examine what determines the shear strength of RC members by obtaining data that cannot be obtained through internal stress state experiments.

To investigate the effects of different water-binder ratios ("W/B" hereinafter) on the shear strength of R-HFDFRCC RC beams, we performed loading tests on R-HFDFRCC RC beams of different W/B. Additionally, we performed uniaxial compression tests and trisecting point bending tests and extracted the fracture mechanics parameters that affect the R-HFDFRCC constitutive equations. Additionally, we factored these constitutive equations into the finite element method (FEM) analysis general code and analyzed the plastic deformation behavior of R-HFDFRCC RC beam specimens. For certain factors, we investigated the effects of the presence or absence of shear reinforcement bars on the shear strength of the R-HFDFRCC RC beams.

Table 1. Concrete specimens.

Specimen	Water-binder ratio (%)	Sand-binder ratio (%)	Fiber volume fraction (vol.%)	Replacement ratio of fly ash (%)
R-HFDFRCC-40-56D	40	40	3.0	20
R-HFDFRCC-50-56D	50	65		
R-HFDFRCC-50-56D@150				
R-HFDFRCC-60-56D	60	90		

2. Experiment method

2.1. HDFRCC

TABLE I provides an overview of the R-HFDFRCC used in this study. The recycled fine aggregate (R) is a mixture of medium fine (max. aggregate dimensions: 2.5 mm; surface dry density: 2.57 g/cm3; water absorption rate: 2.96%; fineness modulus: 2.61) and very fine (max. aggregate dimensions: 0.6 mm; surface dry density: 2.55 g/cm^3; water absorption rate: 4.46%; fineness modulus: 1.16). The cement was ordinary Portland cement (density: 3.16 g/cm^3). The R-HFDFRCC W/B values were 40, 50, and 60%. The R-HFDFRCC material test specimens and R-HFDFRCC RC beam specimens were demolded two days after being placed, then wet-cured until the cumulative temperature inside the curing chamber reached 1,680°DD {equivalent to age 56 days (56D)}, at which point testing was done. We used PVA fiber (diameter: 0.2 mm, length: 18 mm, elastic modulus: 27 kN/mm^2, tensile strength: 975 N/mm^2), and the fiber volume mixing ratio (V_f) was 3%. The admixture was high performance AE water reducing agent, separation reducing agent, and type II fly ash (density: 2.30/cm^3; cement replacement ratio: 20%). The title is set 17 point Times Bold, flush left, unjustified. The first letter of the title should be capitalized with the rest in lower case. It should not be indented. Leave 28 mm of space above the title and 10 mm after the title.

2.2. Material test

In this study, we performed R-HFDFRCC uniaxial compression tests, trisecting point bending tests, pull out tests, and reinforcement bar tensile tests to enable strength control of the R-HFDFRCC RC beam specimens and to extract the fracture mechanics parameters that affect the R-HFDFRCC material constitutive equation.

Table 2. Material properties.

(a) R-HFDRCC

Specimen	Compression			Bending			Pull-out	
	Compressive strength F_c (N/mm^2)	Young's modulus E (kN/mm^2)	Compressive fracture energy G_{Fc} (N/mm)	Bending strength (N/mm^2)	Tensile strength F_{tb} (N/mm^2)	Ulitimate tensile strain ($\varepsilon_{tu,b}$)	Bond strength τ_{max} (N/mm^2)	Slip of bond strength S_u (mm)
R-HFDFRCC-40-56D	50.4	19.3	64.7	7.15	2.25	0.0190	17.1	0.205
R-HFDFRCC-50-56D	35.4	15.9	59.7	6.52	2.25	0.0302	15.8	0.357
R-HFDFRCC-50-56D@150	35.7	15.4	55.8	6.27	2.28	0.0263	14.0	0.327
R-HFDFRCC-60-56D	28.6	14.1	53.6	6.34	1.88	0.0330	12.3	0.418

(b) Reinforcement

<table>
<tr><th rowspan="2">Specimen</th><th colspan="3">Main reinforcement</th><th colspan="3">Shear reinforcement</th></tr>
<tr><th>Ratio (%)</th><th>Young's modulus (kN/mm²)</th><th>Yield strength (N/mm²)</th><th>Ratio (%)</th><th>Young's modulus (kN/mm²)</th><th>Yield strength (N/mm²)</th></tr>
<tr><td>R-HFDFRCC-40-56D</td><td rowspan="4">5.88</td><td rowspan="4">197</td><td rowspan="4">511</td><td rowspan="3">—</td><td rowspan="3">—</td><td rowspan="3">—</td></tr>
<tr><td>R-HFDFRCC-50-56D</td></tr>
<tr><td>R-HFDFRCC-60-56D</td></tr>
<tr><td>R-HFDFRCC-50-56D@150</td><td>0.951</td><td>200</td><td>373</td></tr>
</table>

The test specimens were, in the uniaxial compression tests, a 100 × 200 mm cylinder; in the trisecting point bending tests a 100 × 100 × 400 mm prism; in the pull out tests a 100 × 100 × 100 mm prism with a D-16 (SD490) reinforcement bar inserted; and, for the reinforcement bar tensile tests, a D-16 (SD490) rod-like specimen whose parallel part length was 10 times the nominal diameter or greater. The number of specimens produced per test was six for the uniaxial compression tests and the trisecting point bending tests and three for the pull out tests and reinforcement bar tensile tests.

The uniaxial compression tests were carried out by the method described in reference work [4]. The measurement objects were load, longitudinal and transverse strain at the central part of the specimen (as measured by a compressometer), and displacement between loading boards (as measured by a high sensitivity displacement meter). Additionally, the compression fracture energy (G_{Fc}) was calculated by the method described in reference works [4]-[5] (plastic deformity in reference works [4]-[5] up to 3.0 mm).

The trisecting point bending tests were carried out by the method described in reference work [6]. The measurement objects were load, displacement in the span center portion (as measured by a high sensitivity displacement meter), and curvature (as measured by a pi-shaped displacement meter). Tensile strength ($F_{t,b}$) and ultimate tensile strain ($\varepsilon_{tu,b}$) were calculated by the method described in reference work [7], based on the appendix (reference) to reference work [6].

The pull out tests were carried out by the method described in reference work [8]. The measurement objects were load and degree of slip of the reinforcement bars.

The reinforcement bar tensile tests were carried out by the method described in reference work [9]. The measurement objects were load, longitudinal and transverse strain at the central part of the specimen, and degree of elongation of the reinforcement bars.

We gathered measurement data from the material tests using data loggers. TABLE II gives the material properties of the reinforcement bars and R-HFDFRCC obtained from the material tests.

2.3. RC beam load tests

Fig. 1 provides an overview of the R-HFDFRCC RC beam specimens. Fig. 2 provides an overview of the load tests. For the R-HFDFRCC RC beam specimens, with reference to reference works [10]-[11], the main reinforcement bars were D-16 (SD490), tensile reinforcement ratio (P_t)=5.88%, shear reinforcement bar D-10 (SD295A), shear reinforcement ratio (P_w)=0.951%, and specimen beam depth

(D) 180 mm × beam breadth (b) 100 mm × beam length 1,500 mm. The distance between the fulcrums (L) was 1,300 mm; the shear span length (a) 450 mm; and the distance between load points was 400 mm. The main reinforcement bars were welded to a fixed steel plate (thickness 6 mm) at both ends of the specimen. Loading was done using a 1,000 kN universal testing machine. The measurement objects were load, displacement in the span center portion (as measured by a high sensitivity displacement meter) and the main reinforcement bar, and shear reinforcement bar strain (as measured with a strain gauge). We used data loggers to gather measurement data and calculated the shear strength of RC beams of ordinary concrete using the following formula given in reference work [12].

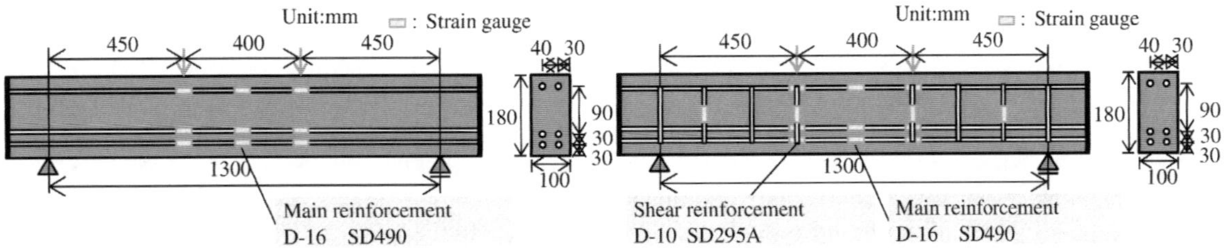

(a) Shear reinforcement (Absence) (b) Shear reinforcement(Presence)

Figure 1. Typical details of relationship.

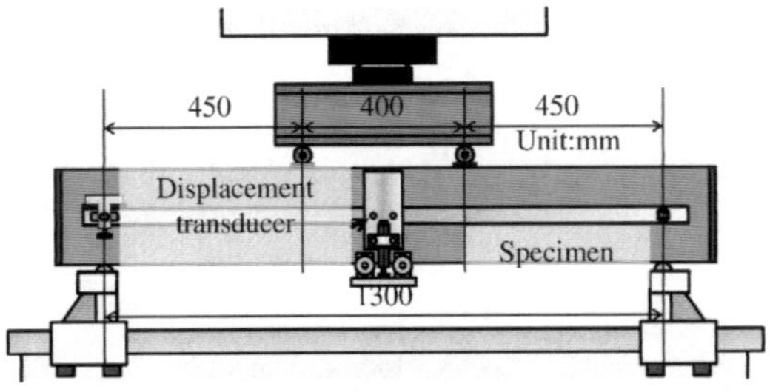

Figure 2. RC beam 4-points bending test.

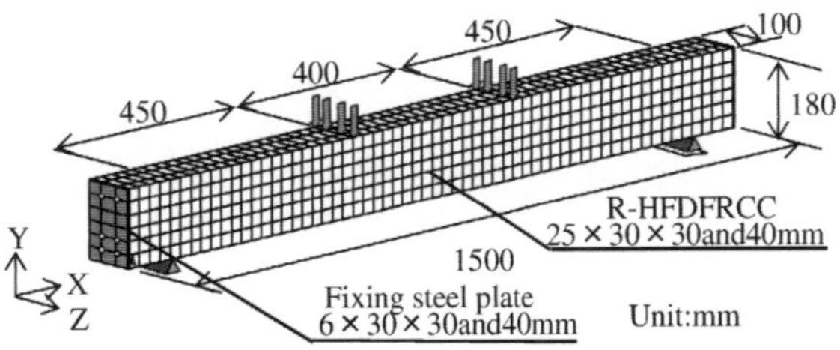

Figure 3. FEM mesh division of RC beam specimen.

$$Q_{bu} = \left\{ \frac{0.092\ k_u k_p (18 + F_c)}{M/(Qd) + 0.12} + 0.85\sqrt{P_w \sigma_{wy}} \right\} bj \tag{1}$$

Here, Q_{bu} is the shear strength (N); k_u: compensation factor 1.0, according to the cross-sectional dimension of the beam; k_p: compensation factor $0.82P_t^{0.23}$ according to the tensile reinforcement ratio; F_c: concrete compressive strength (N/mm^2); $M/(Q_d)$: shear span ratio; P_w: shear reinforcement ratio (%); σ_{wy}: shear reinforcement bar yield strength (N/mm^2); d: effective beam depth (mm); b: beam breadth (mm); and j: distance between stress centers (mm).

3. Outline of analysis

3.1. Methods for modeling and analyzing specimens

In this study, we performed 3D nonlinear FEM analysis of loading tests for R-HFDFRCC RC beam specimens. The R-HFDFRCC RC beam specimens were divided into elements of the dimensions 25 × 30 × 30 mm and 40 mm for the R-HFDFRCC and 6 × 30 × 30 mm and 40 mm for the fixed steel plates. Fig. 3 shows the division of the elements of the R-HFDFRCC RC beam specimens. Each element is an eight-node isoparametric element; the main reinforcement bars are bond-stress slip embedded reinforcement elements. The reinforcement bars were modeled as truss elements. The solid elements and the truss elements were bonded by the interface elements [13].

With respect to the analysis method, a dead weight was initially added, after which a forced displacement was added incrementally at the points indicated by arrows in Fig. 3. For the analysis code, we used DIANA 9.4.4 [13], a general structural analysis program. We applied the Newton-Raphston method as the nonlinear iterative calculation method.

3.2. Material constitutive equation

In response to R-HFDFRCC fracturing, we established an equation based on the total strain on the compressive and tension sides and assumed cracking would take place according to a distributed cracking model that accounted for crack rotation.

The parabola of Fig. 4 gives the stress-strain relationship on the R-HFDFRCC compression side. The area surrounded by the stress descending zone is considered to be a G_{Fc}/representative length (L_c) of elements. G_{Fc} is the material test result shown in Fig. 2, while L_c is the diameter of a sphere having the same volume as the element volume. We also considered the increase in compressive strength attributable to lateral restraints proposed by Vecchio [15] et al. and the reduction in compressive strength of cracked concrete proposed by Collins [16] et al.

A multi-linear model proposed by one of the authors [8] was applied to the stress–strain relationship on the R-HFDFRCC tension side (Fig. 5).

We applied a multilinear model between the R-HFDFRCC and beam main reinforcement bars, constructed from the bond-stress slip relationship obtained in the pull test. The bond-stress at each point on the bond-stress slip model was taken as the average value of the experiment results. Additionally, in consideration of the bond stress slip relationship obtained from the experiments, the slip at each point was taken to be 1/20 of the bond strength slip (S_u) at the first point, S_u at the second point, 2.5 times S_u at the third point, and 7.5 times the third point at the fourth point. Fig. 6 shows the bond-stress slip relationship applied between the beam main reinforcement bars and R-HFDFRCC.

We applied the von Mises yield criterion as the yield criterion for the reinforcement bars. In

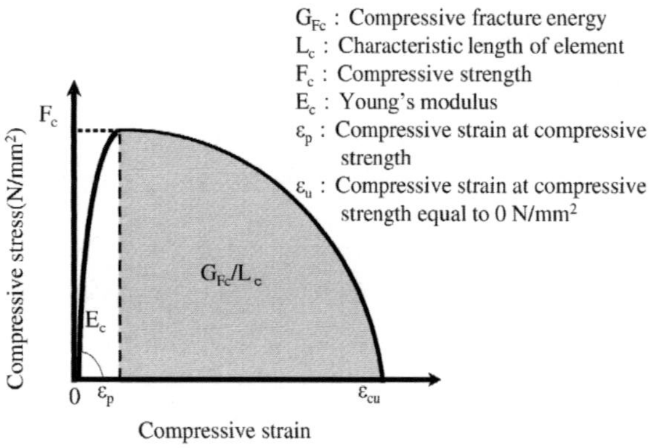

Figure 4. Compression model.

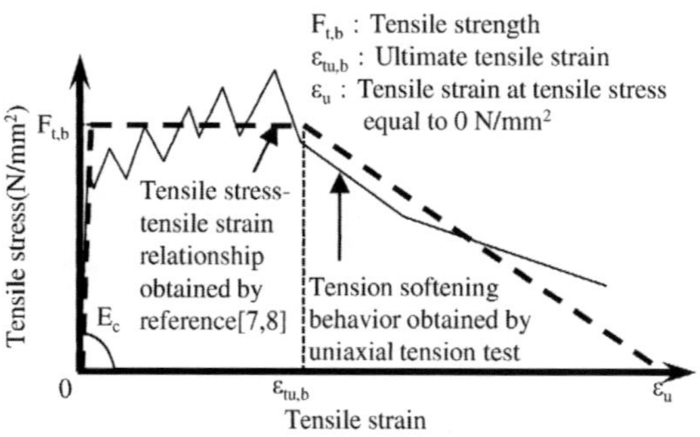

Figure 5. Tension model.

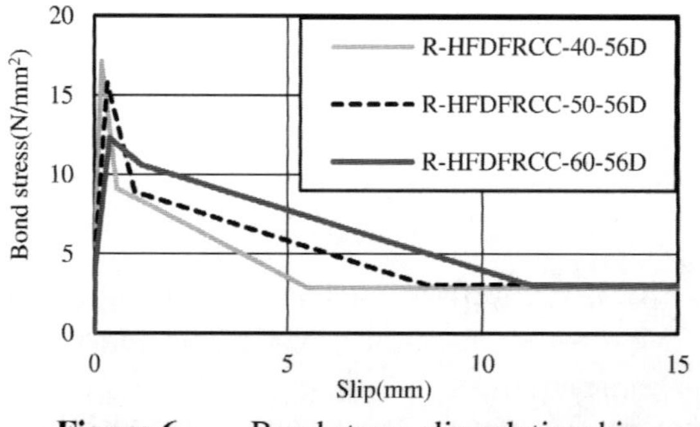

Figure 6. Bond stress-slip relationship.

accordance with the reinforcement bar yield strength and Young's modulus (TABLE II), we formulated a bilinear model of the stress-strain relationship. Additionally, the rigidity of the second gradient was 1/100 of Young's modulus.

4. Results and observations

TABLE III shows the maximum loads obtained from the R-HFDFRCC RC beam specimen loading tests and analyses.

4.1. HFDFRCC RC beam specimen load-displacement relationship

Fig. 7 shows the load-displacement relationship obtained from the R-HFDFRCC RC beam specimen loading tests. The triangle symbols superimposed on each result in the fig show the maximum loads (the same for Fig. 8). In tests without shear reinforcement bars, the maximum load was reached without the main reinforcement bar yielding, terminating in shear fracture. In tests with shear reinforcement bars, the maximum load was reached after the main reinforcement bars had yielded, resulting in fracture.

Looking at the influence of W/B {Fig. 7 (a) and TABLE III}, we see that the maximum load of R-HFDFRCC-60-56D fell by 2.88% compared to the maximum load of R-HFDFRCC-50-56D. However, although R-HFDFRCC-40-56D has a lower W/B than R-HFDFRCC-50-56D, the maximum load of R-HFDFRCC-40-56D fell by 1.80% compared to the maximum load of R-HFDFRCC-50-56D. In TABLE II above, the F_c obtained from the strength control specimen is R-HFDFRCC-40-56D > R-HFDFRCC-50-56D. Here, if the R-HFDFRCC F_c (in TABLE II above) obtained in the material test is substituted into formula (1) used to calculate the shear strength of RC beams using ordinary concrete and the calculations of shear stress performed, the calculation shear strength for R-HFDFRCC-40-56D is 58.7 kN, and that of R-HFDFRCC-50-56D 45.9 kN. According to reference [10], when calculating the shear strength of RC members made of multiple fine crack type fiber reinforced cementitious

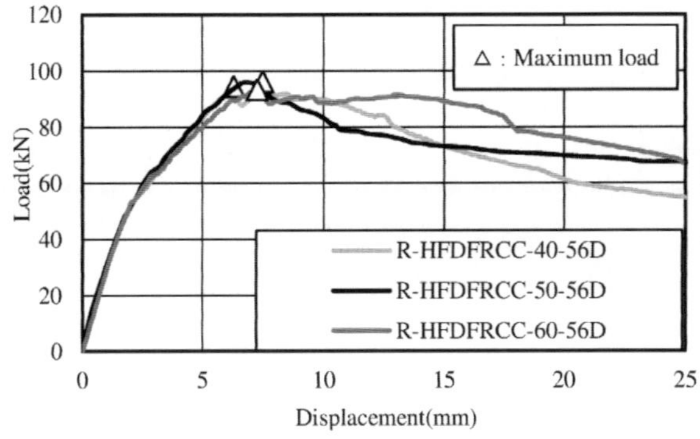

(a) The influence of W/B

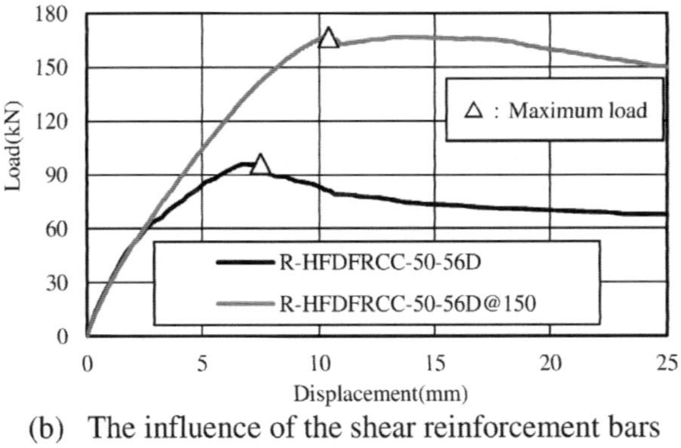

(b) The influence of the shear reinforcement bars

Figure 7. Load-displacement relationship (Experiment).

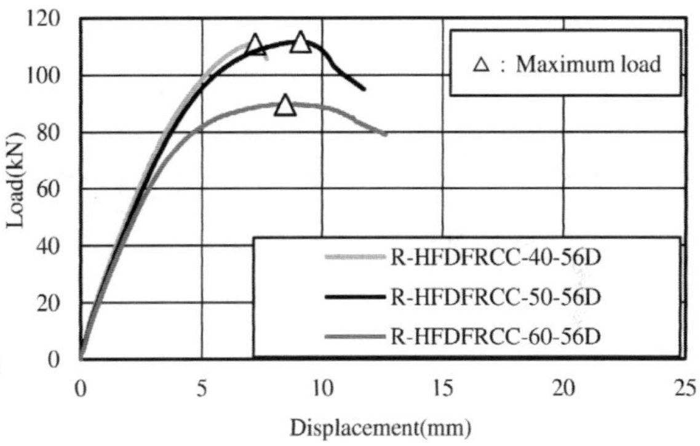

Figure8. Load-Displacement relationship (Analysis).

Table 3. The maximum loads obtained from R-HFDFRCC RC beam specimen the loading tests and analysis

Specimen	Maximum load(kN)	
	Experiment	Analysis
R-HFDFRCC-40-56D	94.6	111
R-HFDFRCC-50-56D	96.3	112
R-HFDFRCC-50-56D@150	167	
R-HFDFRCC-60-56D	93.6	89.8

composites, we must account for the design tensile yield strength of the multiple fine crack type fiber reinforced cementitious composite (i.e., the condition that the average value of the ultimate tensile strain obtained from the uniaxial test is 0.5% or more). According to TABLE II above, the $F_{t,b}$ of R-HFDFRCC-40-56D and R-HFDFRCC-50-56D are equivalent. However, $\varepsilon_{tu,b}$ obtained from the strength control specimen is R-HFDFRCC-40-56D < R-HFDFRCC-50-56D. In addition to F_c and $F_{t,b,}$ the tensile toughness of R-HFDFRCC appears to have affected the maximum load of the R-HFDFRCC RC beam.

Looking at the influence of the shear reinforcement bars {Fig. 7(b) and TABLE III}, we see that the maximum load of R-HFDFRCC-50-56D@150 has increased 73.4% against that of R-HFDFRCC-50-56D. As mentioned above, the fracture of R-HFDFRCC-50-56D@150 with shear reinforcement bars is bending fracture. Under these experimental conditions, we found that making P_w 0.951% caused the shear fracture to change to bending fracture.

Fig. 8 shows the load-displacement relationship obtained by analyzing the loading test R-HFDFRCC RC beam specimens. Additionally, all analysis results confirmed that the maximum load was reached without the main reinforcement bars yielding, and the experimental fracture mode proved reproducible. In considering the effects of W/B {Fig. 8 and TABLE III} compared to the maximum loads obtained from the experiments described above, we see a strong influence with R-HFDFRCC-40-56D. This is slightly lower with R-HFDFRCC-60-56D. However, this analysis also evaluated the tendency to decline of the maximum load of R-HFDFRCC-40-56D compared to that of R-HFDFRCC-50-56D.

4.2. The effects of various material properties on the maximum load of R-HFDFRCC RC beam specimens

Based on IV A, above, we analyzed the effects of F_c, G_{Fc}, $F_{t,b}$, $\varepsilon_{tu,b}$, bond strength (τ_{max}), and bond strength slip (S_u) on the maximum load of the R-HFDFRCC RC beam specimens. TABLE IV gives the analysis factors. In the analysis, taking the analysis of the loading test of R-HFDFRCC RC beam specimens (R-HFDFRCC-50-56D) as the reference criterion, we assumed as parameters F_c, G_{Fc}, $F_{t,b}$, $\varepsilon_{tu,b}$, τ_{max}, and S_u. Additionally, the fluctuation ranges of the parameters were set with reference to the material properties of R-HFDFRCC with different W/B, as shown in TABLE II above.

Table 4. Analysis factors.

	Compression			Bending		Pull-out	
Specimen	Compressive strength F_c (N/mm^2)	Young's modulus E (kN/mm^2)	Compressive fracture energy G_{Fc} (N/mm)	Tensile strength $F_{t,b}$ (N/mm^2)	Ulitimate tensile strain $\varepsilon_{tu,b}$	Bond strength τ_{max} (N/mm^2)	Slip of bond strength S_u (mm)
R-HFDFRCC-50-56D	35.4	15.9	59.7	2.25	0.0302	15.8	0.357
Case-1	20.0	15.9	59.7	2.25	0.0302	15.8	0.357
Case-2	50.4	15.9	59.7	2.25	0.0302	15.8	0.357
Case-3	35.4	15.9	55.0	2.25	0.0302	15.8	0.357
Case-4	35.4	15.9	65.0	2.25	0.0302	15.8	0.357
Case-5	35.4	15.9	59.7	2.00	0.0302	15.8	0.357
Case-6	35.4	15.9	59.7	2.50	0.0302	15.8	0.357
Case-7	35.4	15.9	59.7	2.25	0.0190	15.8	0.357
Case-8	35.4	15.9	59.7	2.25	0.0604	15.8	0.357
Case-9	35.4	15.9	59.7	2.25	0.0302	10.0	0.357
Case-10	35.4	15.9	59.7	2.25	0.0302	20.0	0.357
Case-11	35.4	15.9	59.7	2.25	0.0302	15.8	0.200
Case-12	35.4	15.9	59.7	2.25	0.0302	15.8	0.500
Case-13	50.4	15.9	59.7	2.25	0.0190	15.8	0.357
Case-14	50.4	15.9	59.7	2.25	0.0604	15.8	0.357

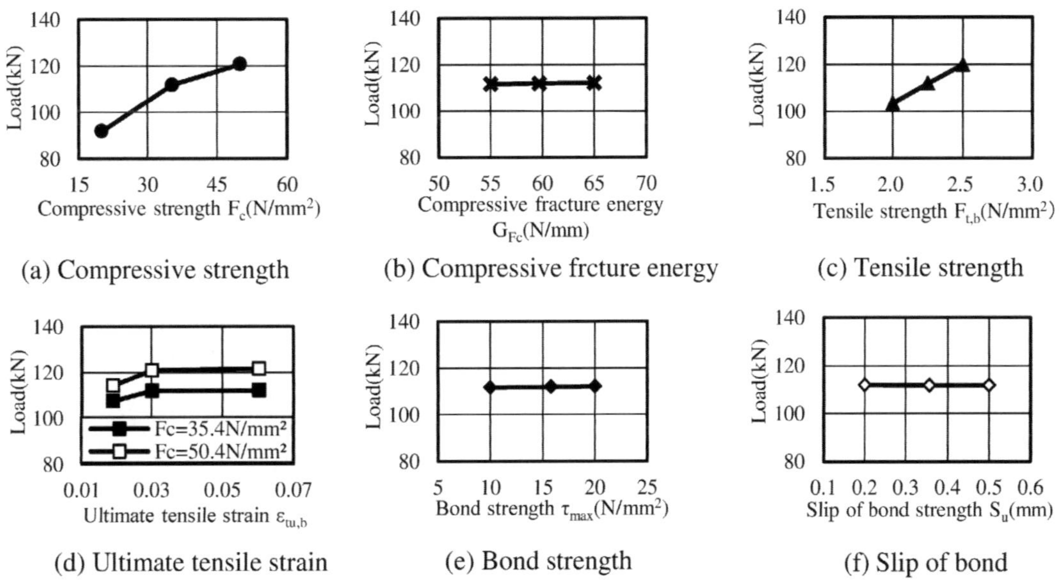

Figure 9. The relationships between the maximum loads obtained from the analyses and analytical parameters.

Fig. 9 shows the relationships between the maximum loads obtained from the analyses and analytical parameters.
According to Fig. 9 (a) (Case 1, R-HFDFRCC-50-56D and Case 2), when F_c increased to 20.0 between 35.4 N/mm^2, the maximum load increased by 20.1 kN; when F_c increased from 35.4 to 50.4 N/mm^2, the maximum load increased by 9.08 kN.
According to Fig. 9 (b), 9 (e), and 9 (f), when G_{Fc} is in the range between 55.0 to 65.0 N/mm, τ_{max} between 10.0 to 20.0 N/mm^2, and S_u between 0.200 to 0.500 mm, the maximum load does not fluctuate significantly, regardless of the differences between G_{Fc}, τ_{max}, and S_u.
Next, according to Fig. 9 (c) (Case 5, R-HFDFRCC-50-56D and Case 6), when $F_{t,b}$ is in the range between 2.00 to 2.50 N/mm^2, the maximum load increases linearly with the increase in $F_{t,b}$. When $F_{t,b}$ increases from 2.00 to 2.50 N/mm^2, the maximum load increases by 16.4 kN.
According to the results in the event of Fig. 9 (d), where F_c = 35.4 N/mm^2 (Case 7, R-HFDFRCC-50-56D and Case 8), when $\varepsilon_{tu,b}$ increases from 0.0190 to 0.0302, the maximum load increases by 4.45 kN. According to the results when F_c = 50.4 N/mm^2 (Case 13, Case 2, and Case 14), when $\varepsilon_{tu,b}$ increased from 0.0190 to 0.0302, the maximum load increased by 6.86 kN. It is understood that the effects of $\varepsilon_{tu,b}$ on maximum load differs depending on the reference F_c. However, even if $\varepsilon_{tu,b}$ is magnified from 0. 0302 to 0.0604, regardless of the difference in F_c, the maximum load hardly increases.
Based on the foregoing and within the scope of this study, the R-HFDFRCC material properties that have significant influence on the maximum loads of R-HFDFRCC RC beam specimens are F_c and $F_{t,b}$. Additionally, the $\varepsilon_{tu,b}$ of R-HFDFRCC, within the range of 0.0190 to 0.0302, affects the maximum load of the R-HFDFRCC RC beam specimens.
In short, the fluctuations in the maximum load of the RC beam specimens in accordance with differences in W/B can generally be explained analytically as long as we understand the differences in the material properties (mainly compressive strength, tensile strength, and ultimate tensile strain) of R-HFDFRCC.
As further analysis, we performed analysis with Young's modulus (E_c) and Poisson's ratio (ν) as parameters, with the R-HFDFRCC RC beam specimen (R-HFDFRCC-50-56D) loading test serving as the reference criterion. When E_c was increased from 15.9 to 19.3 kN/mm^2, we found the maximum load decreased by 1.30 kN; when ν increased from 0.207 to 0.229, the maximum load decreased by 2.58 kN, indicating that these values also affected maximum load to some degree.

5. Conclusion

We obtained the following findings within the scope of this study:

1) The maximum load of an R-HFDFRCC RC beam specimen with a water-binding ratio of 40% decreases compared to that with a water-binding ratio of 50%. Analysis confirmed this tendency.
2) When the water-binding ratio is 50%, the maximum load of an R-HFDFRCC RC beam specimen with shear reinforcement bars increased 73.4% compared to that of an R-HFDFRCC RC beam specimen without shear reinforcement bars.
3) The maximum load of an R-HFDFRCC RC beam specimen is affected within an R-HFDFRCC ultimate tensile strain range of 0.0190 to 0.0302.
4) The fluctuations in the maximum load of R-HFDFRCC RC beam specimens in accordance with differences in water-binding ratios can generally be explained as long as the differences in the material properties (mainly compressive strength, tensile strength, and ultimate tensile strain) of R-HFDFRCC are understood.

6. References

[1] JCI, Technical Committee on High Performance Fiber Reinforced Cementitious Composites, Committee Report on High Performance Fiber Reinforced Cementitious Composites, The Japan Concrete Institute , Japanese: 2002, pp.1-10

[2] JCI, Technical Committee on Utilization of High-Strength and High-performance Concrete, Committee Report on Utilization of High-strength and High-performance Concrete, The Japan Concrete Institute , Japanese: 2009, pp.74-85

[3] Ohtsu, N and Watanabe, K, "Fundamental Study on High-Fluidity Ductile-Fiber-Reinforced Cementitious Using and Recycled Fine Aggregate," Proceedings of the Japan Concrete Institute, 38(1), pp.357-362, Japanese, 2016.7

[4] Watanabe, K, Ohoka, T, Hakuto, and Kato, Y, "Compressive Fracture Behavior of DFRCC Using Recycled Fine Aggregate, " Proceedings of the Japan Concrete Institute, 28(1), pp.485-490, Japanese, 2006.7

[5] Watanabe, K, Ohoka, T, Hakuto, Shirai, N ,and Morizumi, K , "Compression Softening Behavior of Various Type of Concrete, "Proceedings of the Japan Concrete Institute, 22(2) , pp.493-498, Japanese, 2000.6

[6] JCI, Method of Test for Bending Moment-Curvature Curve of Fiber-Reinforced Cementitious Composites JCI-S-003-2007, The Japan Concrete Institute, Japanese: 2007, 8pp.

[7] Watanabe, K, Sato F , Miura Y and Shibuya, K, "Tension Softening Behavior of DFRCC Using Various Type of Fine Aggregate, " Proceedings of the Japan Concrete Institute, 32(1) pp.287 - 292, Japanese, 2010.7

[8] JTCCM, Method of Test for Bond Strength Between Concrete and Steel Reinforcement by Pull Out Test JSTM C 2101 2005, Japan Testing Center for Construction Materials Standard, Japanese:2005

[9] JIS, Metallic Materials-Tensile Testing-Method of Test at Room Temperature (JIS Z 2241), Japanese:2011

[10] JSCE, Recommendations for Design and Construction of High Performance Fiber Reinforced Cement Composite with Multiple Fine Cracks, Japan Society of Civil Engineers, Japanese: pp.30-32

[11] Nagai, S, Takaine, Y, Kanda, T and Maruta, M, "Shear and Flexural Performance of Reinforced Beam Using High Performance Fiber Reinforced Cement Composite : Part2 Considerations of the Test Result, " Summaries of Technical Papers of Annual Meeting Architectural Institute of Japan pp.313-314, Japanese, 2001.9

[12] AIJ, AIJ Standard for Standard for Structural Calculation of Reinforced Concrete Structures – Based on Allowable Stress Concept– revised 2010, Japanese:2010, 525pp

[13] TNO, DIANA Foundation Expertise Center for Computational Mechanics (DIANA version-9) : DIANA Finite Element Analysis Users Manual, TNO Building and Construction Research. 2003

[14] Selby, R.G, and Vecchio, FJ, "Three-dimensional Constitutive Relations for Reinforced Concrete. Tech. " Univ. Toront : 1993, Rep. 93-02,

[15] Vecchio, FJ and Collins, "Compression Response of Cracked Reinforced Concrete, " pp.3590-3610, ASCE, 1993As part of the production system for *IOP Conference Series: Materials Science and Engineering*, online versions of all reference lists will, wherever possible, be linked electronically using CrossRef.

Acknowledgments
A portion of this research was undertaken with the aid of a JSPS Grant-in-Aid for Scientific Research (Issue Number: 15K06307, Representative: Ken Watanabe).

IOP Conf. Series: Materials Science and Engineering **264** (2017) 012010 doi:10.1088/1757-899X/264/1/012010

Verification of chloride adsorption effect of mortar with salt adsorbent

T Hoshina[1], N Nakajima[2], H Sudo[3] and S Date[4]

1 Tokai University/Department of Civil Engineering, Hiratsuka City, Japan
2 Tokai University/Department of Civil Engineering, Hiratsuka City, Japan
3 Tokai University/Department of Civil Engineering, Hiratsuka City, Japan
4 Tokai University/Department of Civil Engineering, Hiratsuka City, Japan
E-mail: takuyahoshina117@gmail.com

Abstract. In order to investigate the chloride adsorption effect of mortar mixed with chloride adsorbent, electrophoresis test using mortar specimen and immersion dry repeated test were conducted to evaluate chloride adsorption effect. As a result, it was confirmed that soluble salt content that causes corrosion of rebar in the specimen was reduced by the chloride adsorbent and corrosion inhibiting effect of the rebar was also obtained. It was also confirmed that by increasing dosage of the chloride adsorbent, the chloride adsorbing effect becomes larger as well.

1. Introduction

Recently, deterioration of the RC structure caused by corrosion of the rebar due to penetration of chloride ion is being concerned. Salt damage often causes fatal cracking and peeling of concrete member due to expansion of corroded rebar. Ground structure under construction in the high economic growth period, especially common precast products such as RC segments for bridge on the coast are aged and intruded chloride ion, as the result, deterioration by the salt damage can be observed. Therefore, countermeasures against salt damage deterioration are being required and developed restorative materials mixed with salt adsorbent [1]-[5] .

At the same time, a new durable RC structure capable of suppressing salt damage deterioration is being desired for extension of lifetime of the RC structure. [6]-[12]Salt content which is a deterioration factor of the RC structure penetrates from the surface through cover portion which is the surface layer. If it is possible to immobilize the infiltrated soluble salt in the cement hydrate, it is considered that deterioration of the infrastructure by salt damage can be suppressed In this research, focusing on the cover zone of the newly constructed RC structure, the electrophoresis test and a dry-immersion repetition test to confirm salt adsorption effect by salt adsorbent in the mortar were conducted respectively, furthermore, under severe salt damage deterioration environment, the corrosion of rebar in two tests which verified by self-potential and polarization resistance value were investigated.

2. Experiment summary

2.1. Materials used

The chloride adsorbent used in this study is a high performance salt adsorbent which is sort of layered hydroxide of magnesium - aluminum. High performance chloride adsorbent (hereinafter referred to as HPCA) has the ability of intercalation to capture an anion by layers and release retained a cation. In this study, in order to improve the adsorption and exchange performance, nano-sized HPCA of which crystallite size is around 10 nm, and nitrate ion $(NO3)^-$ retained in the interlayer to prevent corrosion of rebar, and nitric acid type HPCA which simultaneously releases nitrate ions was used as a salt adsorbent. Figure 1 shows the image of chloride ion (Cl -) adsorption and nitrate ion (NO3 -) release of the salt adsorbent.

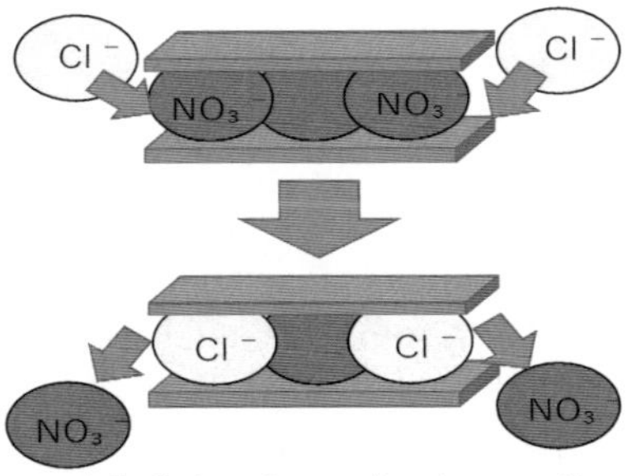

Figure 1.Salinity absorption image diagram

2.2. Compounding condition

For the specimens used for the electrophoresis test, ordinary portland cement, Toyoura standard sand were used. In addition, water cement ratio of the mortal was controled to 50%. The specimens were cut to 50 mm in thickness after 14 days of sealing curing and sealed on the circumferential surface other than the test surface was used.

HPCA was replaced by volume of sand by 20% of the mass of cement. Table 1 shows the mix proportions of specimens. Table 2 shows the sample used in the dry-immersion repetition test in Table 2. Table 3 shows the formulation of the mortar in the covered part. The salt adsorbent was replaced by the volume of the fine aggregate and added 0, 5, and 10% at the end.The outline of the specimen is shown in Fig. The red part is a mortar mixed with a salt adsorbent. Epoxy resin was applied except for the cover surface. Therefore, chloride ions were adjusted so as to permeate only from the fog surface

Table 1: Mortar mixing method

Type of specimen	W/C (%)	Unit weight (kg/m^3)			
		W	C	S	HPCA
N-0%	50	303	605	1210	0
N-20%				1072	121

Table 2: Materials used

Material	symbol	Type	Density (g/cm^3)
Cement	C	Ordinary Portland cement	3.15
Fine aggregate	S1	Land sand from the Oikawa	2.59
HPCA	S2	High performance chloride absorbent	2.3
Coarse aggregate	G	Crushed stone from Sagamihara	2.67
Admixture	Ad	AE water reducing agent	-

IOP Conf. Series: Materials Science and Engineering **264** (2017) 012010 doi:10.1088/1757-899X/264/1/012010

Table 3: Mortar mixing method

	W/C	Unit weight (kg/m^3)				
	(%)	W	C	S1	S2	Ad
H0	40	275	689	1194	0	C×0.23
H5				1155	34	
H10				1116	69	

2.3. Experimental condition

Exam. 1

In the electrophoresis test, a specimen with a diameter of 100 mm and a thickness of 50 mm was placed between the electrodes using a 0.5 mol / L sodium chloride aqueous solution on the cathode side and 0.3 mol / L sodium hydroxide aqueous solution on the anode side, and a direct current constant voltage of 15 V was applied. The outline of the test is shown in Figure 3. To verify the salt adsorption effect of HPCA, each solution at both poles was not exchanged. Afterwards, the solution was collected and measured the chloride ion concentration.

Exam. 2

During dry-immersion repetition test, penetration of chloride ion into specimen used was observed by soaking the specimens in a 3% brine solution at a temperature of 40 ° C and a humidity of 60%, repeating dry and humidity, immersing them for 3 days and 4 days drying was defined as one cycle and natural potential and polarization resistance were measured at each cycle, and a total of 19 cycles were performed.

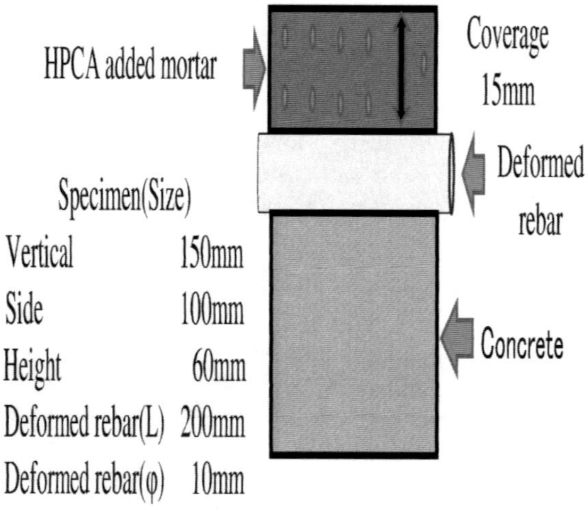

Figure 2: Schematic of the specimen

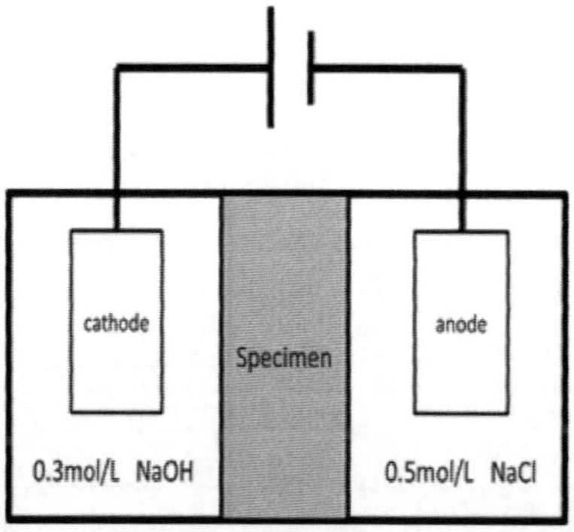

Figure 3: Electrophoresis test summary

IOP Conf. Series: Materials Science and Engineering **264** (2017) 012010 doi:10.1088/1757-899X/264/1/012010

3. Results and discussions

Exam. 1

Changes with time of the chloride ion concentration on the cathode side and the anode side are shown in Figure 4,

Changes over time of the nitrate ion concentration on the anode side are shown in Figure 5 As shown in Figure 4, in compare with the control, it was confirmed that specimen with HPCA were effective to decrease amount of chloride ion, but from the result of Figure 5, it was also confirmed that the increase of chloride ion in HPCA mixed specimens was smaller than control.

Exam. 2

Figure 6 shows the polarization resistance of each specimen, and Figure 7 shows the natural potential of each specimen. From the result of Figure 6, it was observed that each sample shows the same corrosion tendency until the 12th week.

There was a difference in each specimen from 13 weeks, and corrosion proceeded with H 0 at each subsequent cycle number, but it can be seen that the HPCA-added mortar (H 5, H 10) maintained a relatively natural potential. This phenomenon is presumed to be a result caused by chloride ion adsorption effect by HPCA.

In Figure 7, each specimen did not show in a state of corrosion tendency until week of 12. There is a difference in each sample, and H 0 tends to corrode even if the next cycle number overlaps from week of 13 to week of 18, H5 showed a slightly higher polarization resistance than H0. On the other hand, H10 showed significantly higher polarization resistance value than the other two samples, therefore this indicates that HPCA has good potential of corrosion control effect.

However, in both H 5 and H 10, the polarization resistance value decreased at 19 weeks. This is thought to be due to the effect of salt adsorbent decreasing at week of 19.

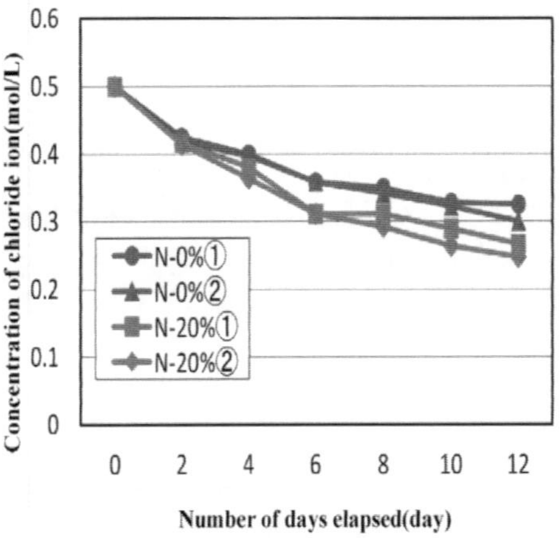

Figure 4: Changes with time of the chloride ion concentration on the cathode side

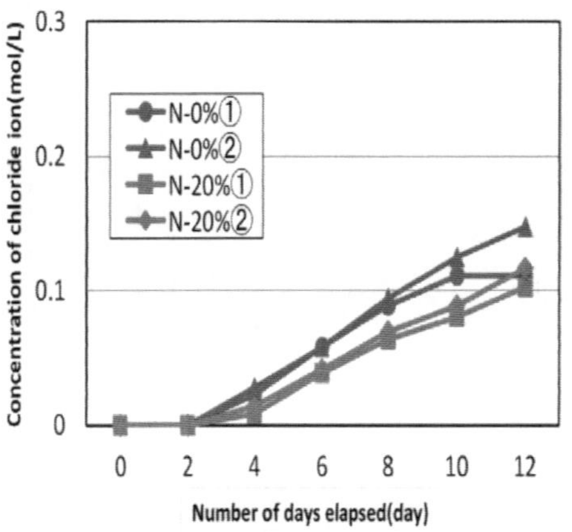

Figure 5: Changes with time of the chloride ion concentration on the anode side

4. Conclusion

Chloride adsorption effect of mortar with HPCA which is chloride adsorbent was verified by the electrophoresis test.

As a result, and it was verified that the penetration of the chloride ion was suppressed.

Dry-immersion repetition test was conducted to verify the effect of chloride adsorption under severe salt damage deterioration environment. As a result, it can be confirmed from the natural potential that the corrosion of rebar is alleviated by adding HPCA to the cover layer of the mortar. From the polarization resistance value at the cover layer of the morta, it is verified that when the salt adsorbent is added by 10% of the cement amount, the effect increases.

By the two tests, it is considered that addition of HPCA to the mortar prevents chloride ions infiltration from the surface to the depth of rebar embedded and adsorbs chloride ion and releases

IOP Conf. Series: Materials Science and Engineering **264** (2017) 012010 doi:10.1088/1757-899X/264/1/012010

corrosion inhibiter to suppression of rebar corrosion. Furthermore, tt was verified that adsorption effect of HPCA was greatly improved as the dosage of HPCA increases.

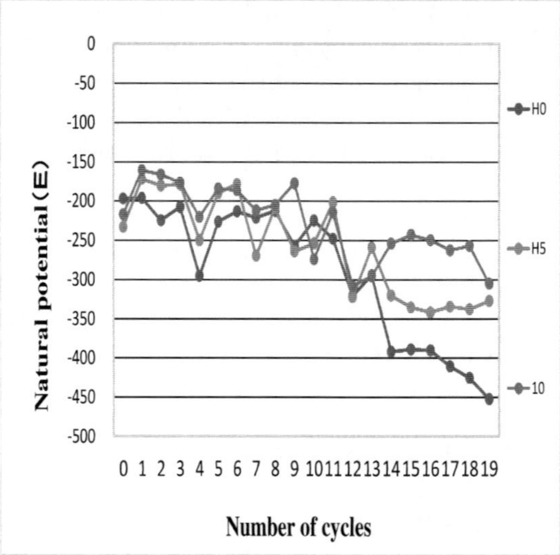

Figure 6. Natural potential

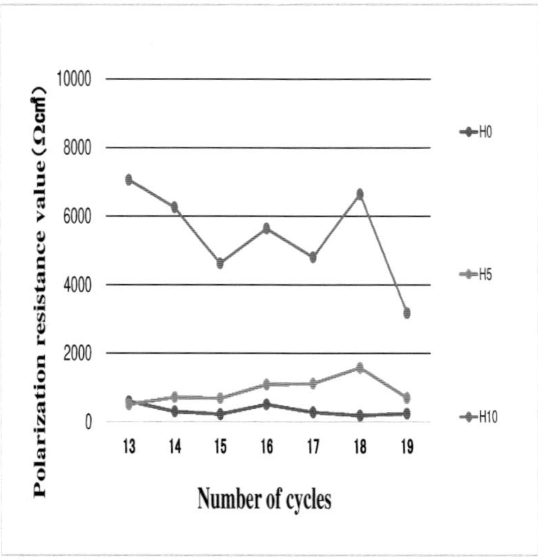

Figure 7. Polarization resistance value

5. References

[1] Pitiwat WATTANACHAI, Nobuaki OTSUKI, Tsuyoshi SAITO and Takahiro NISHIDA:A STUDY ON CHLORIDE ION DIFFUSIVITY OF POROUS AGGREGATE CONCRETES AND IMPROVEMENT METHOD, Dodoku Gakkai Ronbunshuu E, Vol.65, No.1, 30-44, 2009.1

[2] Sung-Ho TAE and Takumi UJIRO: A Study on the Corrosion Resistance of Cr-bearing Rebar in Mortar in Corrosive Environments Involving Chloride Attack and Carbonation, ISIJ International, Vol.47, No.5, 715-722, 2007

[3] Masatoshi KATAOKA, Shouki YATSUSHIRO, Shouhei YAMAMURA, Masato TANAKA and Toshihiro OOIE: Biomarker analysis on maicrochips-Development of POCT device for multi-marker analysis-, Translation from synthesiology, Vol.3, No.1, 16-25, 2010

[4] Sung-Ho TAE, Takafumi NOGUCHI and Takumi UJIRO: Corrosion Inhibition by Cr-bearing Rebar in Concrete Due to Combined Deterioration of Carbonation and Chloride Attack, ISIJ

International, Vol. 47, No. 1, 146–150, 2007

[5] Sung-Ho TAE, Han-Seung LEE, Takafumi NOGUCHI, Takumi UJIRO and Sung-Woo SHIN: Corrosion Resistance of Cr-bearing Rebar to Macrocell Corrosion Caused by Concrete Cover Crack, ISIJ International, Vol. 46, No. 7, 1086–1092, 2006

[6] Yasushi TANAKA, Tetsuya ISHIDA, Ichiro IWAKI and Kazunori SATO: MULTIPLE PROTECTION DESIGN FOR DURABLE CONCRETE BRIDGE DECK IN COLD REGIONS, Journal of JSCE, Vo.5, 68-77, 2017

[7] Sho TSUCHIYA, Mikito UEDA and Toshiaki OHTSUKA: Polypyrrole Coating on Zinc for Corrosion Prevention of Zinc-coated Steels, ISIJ International, Vol. 47, No. 1, 151–156, 2007

[8] Takahiro SASAKI, Tohru IIJIMA and Katsuya KOBAYASHI: Predicting Concrete Deterioration due to Chloride Corrosion of Reinforcing Bars, QR of RTRI, Vol.45, No.4, 2004.11

[9] Hidenobu TATEMATSU, Takahiro SASAKI and Tooru IIJIMA: Studies on Diagnosis and Repair for Reinforcing Bar Corrosion by Salt Injury, QR of RTRI, Vol.42, No.1, 2001.3

[10] Masayasu OOTSU, Sunun KEI and Masaru yokota: Nondestructive inspection method for corrosion of reinforcing bars in concrete Research on practical application, Concrete Education Proceedings, Vol.15, No.3, 2004.9

[11] Yutaka KAWANO: Degradation phenomenon of concrete (about salt damage), IIC REVIEW, No.39, 2008.4

[12] Hidenobu TACHIMATSU, Takaiko SASAKI and Jyun TAKADA: Study on diagnosis and suppression of rebar corrosion due to salt damage, Concrete Engineering paper, Vol.11, No,2, 2000.5

Effects of curing conditions and addition of fine blast furnace slag powder on the initial performance of concrete using expanding material

Nozomi Nakajima[1], Kazuki Hirabe[2], Shin-ichiro Hashimoto[3], and Shigeyuki Date[4]

1 Tokai University, Graduate School of Engineering Building Civil Engineering Major, Hiratsuka city, Japan
2 Tokai University, Graduate School of Engineering Building Civil Engineering Major, Hiratsuka city, Japan
3 Fukuoka University/ Faculty of Engineering Department of Social Design Engineering Assistant Professor, Fukuoka city, Japan
4 Tokai University, Faculty of Engineering Department of Civil Engineering Professor, Hiratsuka city, Japan
E-mail: hayabusa_ripobitann_0613@yahoo.co.jp

Abstract. In this study was investigated the effect of steam curing in the case of adding expandable material to powdery high fluidity concrete containing blast furnace slag fine powder on two kinds of expansive materials, ettringite type inflation material and lime type expansion material. As a result, the constraint expansion strain increases with the ettringite type due to the increase in the steam curing temperature. However, it was confirmed that it decreased with the lime type expansion material.

1. Introduction

In recent years, the Ministry of Land, Infrastructure, Transport, and Tourism JAPAN promotes the activities effort to improve the productivity of the construction site. The merit of use of precast concrete products is labor saving and shortening construction period, and it is drawing attention as a solution to the problem in concrete works again. In addition, since precast concrete products are manufactured at quality controlled factories, it can be considered that quality of precast concrete product is higher than that of cast-in-place concrete. [1] However, most concrete descriptions on concrete standard concrete descriptions, and quality control of Precast Concrete products differ from factory to factory in many cases and areas. [2] - [3] Meanwhile, at the Precast Concrete product factory efforts to reduce costs and improve production efficiency are being carried out daily basis, but one of such efforts is to improve form rotation efficiency. In this case, consideration is made to shorten the preliminary time, etc. [4] - [12] When a precast concrete factory is located near a residential area, high flowing concrete may be used in some cases. Compared with ordinary concrete, high flowing concrete is less time to vibration and therefore it is said to be good for noise countermeasure and so on. High fluidity concrete become contain powder such as slag and fly ash. However, since slag and fly ash have large self-shrinkage, cracks often occur in precast concrete products. In addition, precast concrete products are subjected to high temperature steam curing, so thermal stress cracks occur frequently. Therefore, an inflating agent is used as crack suppression. Therefore, attention is expansion materials and the like as concrete materials. [13] However, it is

recognized that the shrinkage distortion increases and the quality deteriorates such as cracking occurs. For this reason, it is necessary to optimize the curing condition, compounding condition, and material used, but as a result, the expanding material may be most than the general usage amount.
However, there are few investigations and studies on the influence of inflating material exerted by the preliminary time, and it is necessary to systemize it to btain efficient curing condition / compounding condition. In addition, there are cases where highly fluidized concrete, such as SCC is being adopted for labor saving, noise control, and high quality. Also, blast furnace slag fine Strain gauge)eing used as the material for high fluidity concrete. There is also a report that expansion of gypsum decreases when the blast furnace slag fine powder and ettringite type expansion material are used in combination and is consumed in the reaction of slag. In g temperature and preliminary time on expanding material for powdered high fluidity concrete containing blast furnace slag fine powder. In addition, from the viewpoint of quality and productivity of precast concrete products, the initial compressive strength and expansion rate, which are important performance index, were also investigated.

2. Experiment products

2.1. Materials
Table 1 shows the material used and its quality. The specific surface area of the blast furnace slag fine powder used was 4000 brains. The expansion material used was two types (Ettringite type and Lime type).

Table 1. Specimen of restraint expansion strain.

Materials	Symbol	Properties	Density (g/mm^3)
Cement	C	Ordinary Portland cement	3.16
Blast furnace slag fine powder	Sg	With plaster	2.89
Expanding material	CaO	Ettringite type	2.93
	Et	Lime type	2.19
Fine aggregate	S	River sand from Yamakita Kanagfawa	2.61
Coarse aggregate	G	Crushed stone from Sagamihara Kanagawa	2.69
Chemical Admixture	Ad	High performance AE water reducing agent	-

2.2. Mix proportions
The mix proportions of the concrete used in the experiment is shown in Table 2. High fluidity concrete with a water binder ratio (W / B) of 33.6%, various expansion materials were applied internally. The blast furnace slag fine powder replaced slag by 50% with respect to cement. The amount of admixture was adjusted so that the slump flow was in the range of 60 ± 10 cm (P × 0.6% constant)

Table 2. Mix proportions.

<table>
<tr><th>No.</th><th>W</th><th>C</th><th>Sg</th><th>CaO</th><th>Et</th><th>S</th><th>G</th></tr>
<tr><td>PL</td><td rowspan="3">168</td><td>250</td><td>250</td><td>-</td><td>-</td><td>764</td><td>822</td></tr>
<tr><td>CaO</td><td rowspan="2">233</td><td rowspan="2">233</td><td>35</td><td>-</td><td>778</td><td>838</td></tr>
<tr><td>Et</td><td>-</td><td>35</td><td>799</td><td>893</td></tr>
</table>

2.3. Curing Parameter
The curing temperature diagram is shown in Fig.1. The preliminary time (pre-curing) was set to three levels, 0.5h, 1.5h, and 2.5h. In addition, the maximum temperature of steam curing was set to 3 levels of 45 ° C, 55 ° C, and 65 ° C. The steam curing was conducted until demolding time (20 h), and thereafter it was kept in an environment of 20 ° C. and 60% relative humidity.

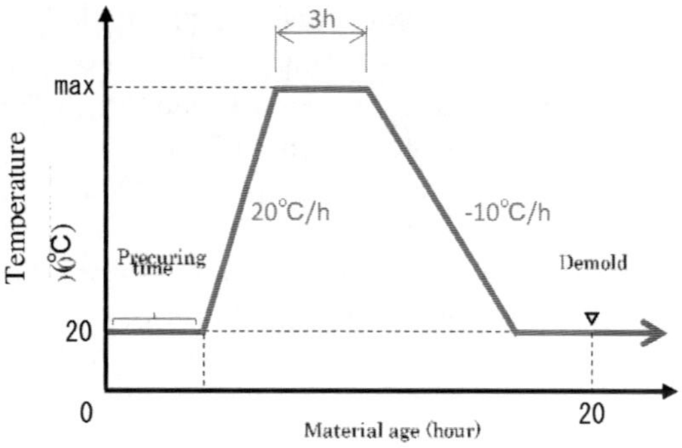

Fig.1. Curing pattern.

2.4. Test Items

Test items were checked for compression strength and length change. The change in length was a constraint expansion strain according to JIS A 6202 B method, and it was measured with the specimen shown in Fig.2. S crew cut steel bars attached two train gauges on the center portion (see Fig.2) were arranged in specimen. Time-dependent change of the strain during steam curing process was also measured. To cancel the thermal expansion of the restraint device, the strain was measured simultaneously in the steam curing chamber at the same time as the dummy restraint device, and it was subtracted from the strain of the concrete specimen.

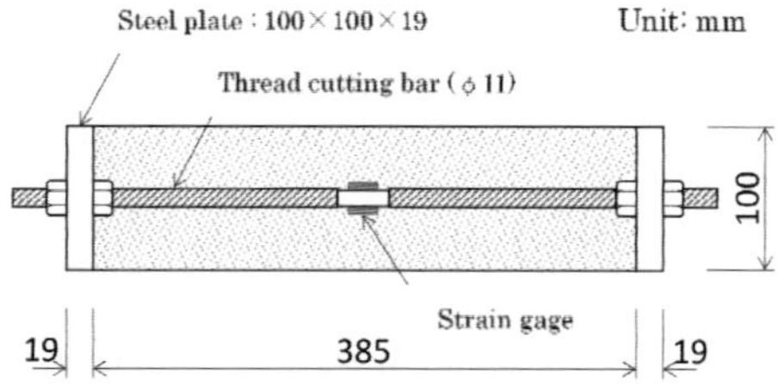

Fig.2. Specimen of restraint expansion strain.

3. Experiment results and discussion

3.1. Compressive strength

Fig. 3 shows the compressive strength test results for each maximum steam curing temperature. First of all, in the case of the maximum temperature of 45 ° C, each compounding showed almost the same performance at the compressive strength (material age 20 h) at the time of demoulding. However, after 14 days of age, the compressive strength was somewhat larger in the case of adding the expandable material. When the maximum temperature is 55 ° C, the strength is enhanced by the addition of the expansion material at the stage of demolding. Even at the maximum temperature of 65 ° C, the same tendency as at 55 ° C was observed, but the strength enhancement at the time of demolding became more prominent. For this reason, in the case of 45 ° C, it is considered that compressive strength (material age 20 h) enhancement appeared after demolding. In the case of precast concrete products, since demolding is one day, it was found that a curing temperature of at least 55 ° C. or higher is required under the curing condition in this study to obtain the effect of enhancing the strength by the expansive material.

Next, Fig.4 shows the relationship between the preliminary time and compressive strength at the time of demolding (material age 20 h). In the PL formulation and the CaO formulation, no significant difference due to the preliminary time was confirmed, but at 45 ° C. and 55 ° C. in the Et formulation, the compressive strength tended to increase as the preliminary time increased. Next, Fig.4 shows the relationship between the preliminary time and compressive strength at the time of demolding (material age 20 h). In the PL formulation and the CaO formulation, no significant difference due to the preliminary time can be confirmed, but at 45 ° C. and 55 ° C. in the Et formulation, the compressive strength tended to increase as the preliminary time increased. It is also possible to be affected by the preliminary time only when using ettringite type expansion agent. However, it was found that there was a possibility that a certain accumulated temperature and an effective material age might be necessary, as this tendency is not seen at 65 ° C and strength enhancement did not appear at 45 ° C during demolding It was.

3.2. Length change test

A length change test result with a maximum temperature of 55 ° C is shown Fig.5. From this result, the expansion reaction due to the expansion material in all cases occurs earlier than the ettringite type (Et) with the lime type (CaO) expansion material and peaks over several hours. The timing was the same for both ettringite and lime, and this tendency was the same even in the case where the maximum temperature was different. In comparison with the maximum expansion strain, the lime type expansion agent is generated about twice for the ettringite type expansion agent, and these have also been confirmed in the past research of the authors. Next, the relationship between the preliminary time at the time of demolding (material age 20 h) and the constraint expansion strain is shown in Fig.6. From this figure, it can be confirmed that the constraint expansion strain tends to be larger in the case where the preliminary time is longer in all the cases. For this reason, it is understood that it is necessary to secure sufficient preliminary time to sufficiently exert the expansion effect of the expansion material irrespective of the curing temperature.

Fig.7. shows the relationship between compressive strength and restraint expansion strain at the time of demolding (material age 20 h). In all cases, the compressive strength of the ettringite (Et) expansion material tends to increase, that is, the restraint expansion strain tends to increase as the maximum temperature of steam curing increases. On the other hand, in the case of the lime type (CaO) expansion material, the constraint expansion strain tended to decrease as the maximum temperature of the steam curing increased. Therefore, in the figure, a broken line is written at a position intermediate between the ettringite type expansion material and the lime type expansion material. As described above, the constraint expansion strain of the both expansion materials of the ettringite-based expansion material and the lime type expansion material is increased as the preliminary time becomes longer. Furthermore, it can also be confirmed that the convergence tends to occur as the compressive strength increases. Compressive strength levels are all equal to the range of 10 to 50 N / mm2. From this, it is also expected that the Young's modulus and the linear expansion coefficient will be different, noting the timing at which the expansion amount appears due to the different preliminary time.

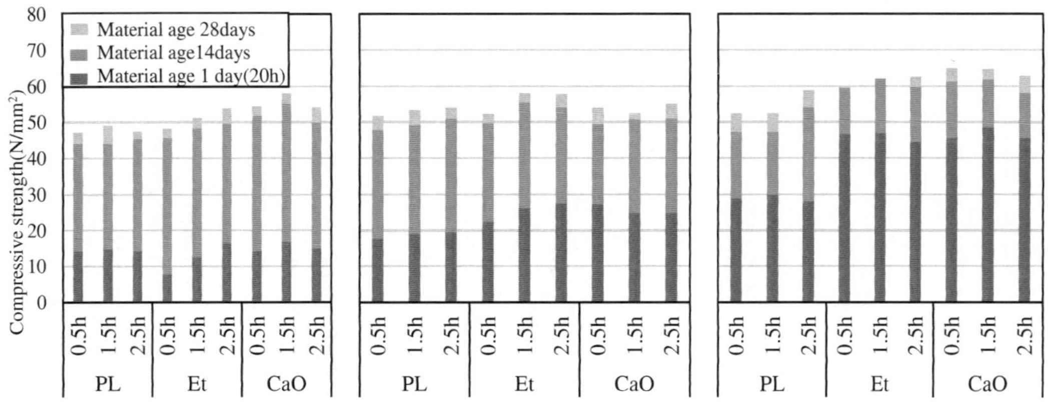

a) Max temparature 45℃ b) Max temparature 55℃ c) Max temparature 65℃

Fig.3. Compressive strength test result.

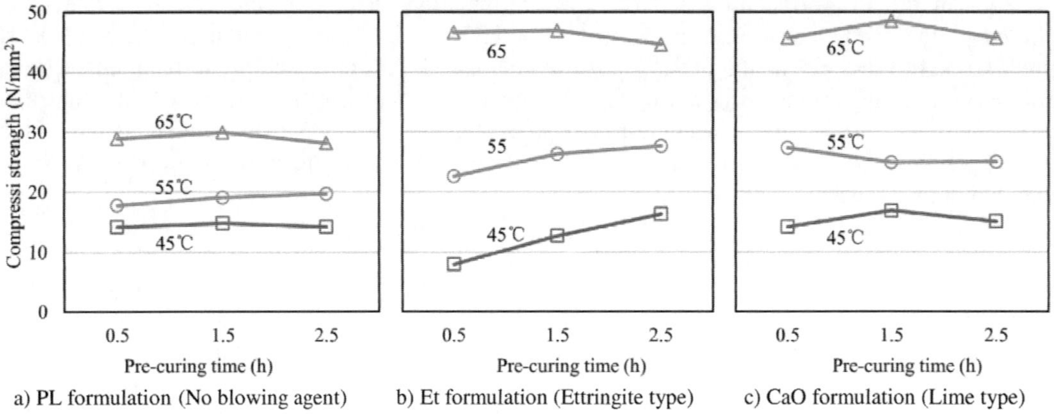

Fig.4. Relationship between preliminary time and compressive strength
At demolding: Material age of 20 hour

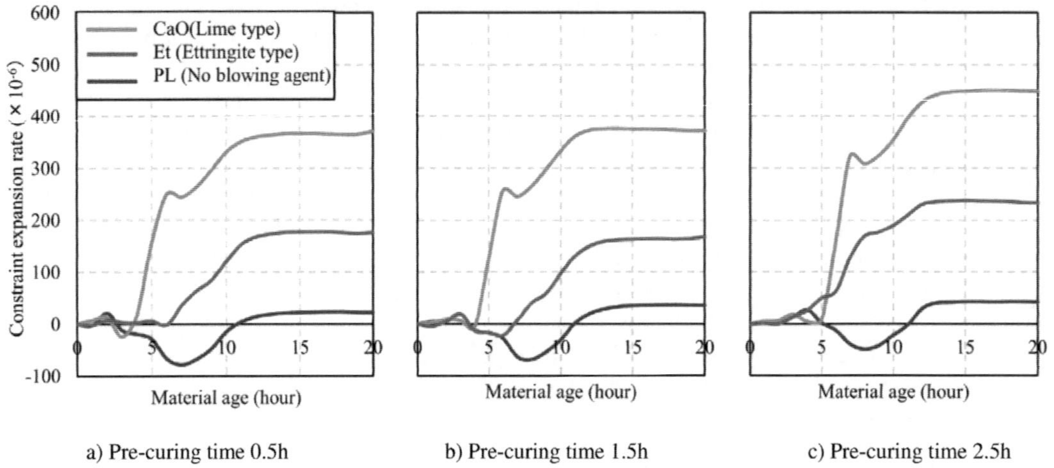

Fig.5. Length change test result (Maximum temperature 55 ° C).

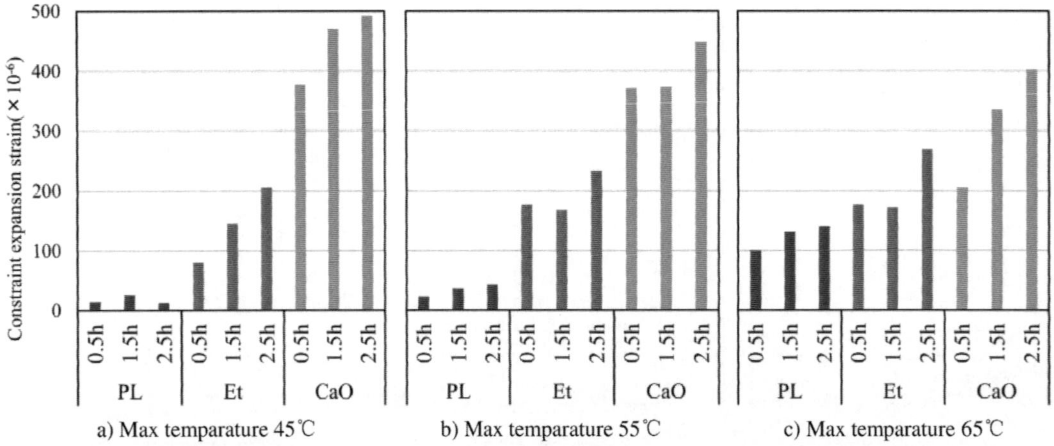

Fig.6. Relationship between Pre-curing time and constraint expansion strain
At demolding: Age of 20 hour.

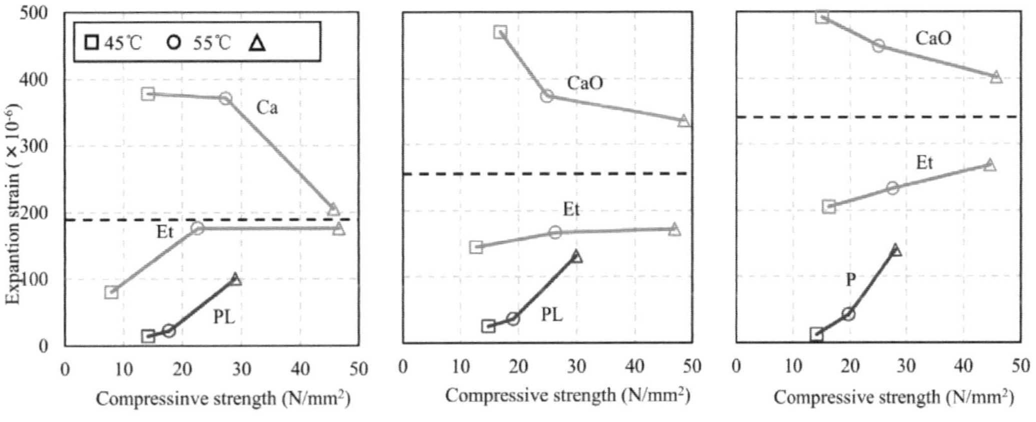

a) Pre-curing time 0.5h b) Pre-curing time 1.5h c) Pre-curing time 2.5h

Fig.7. Relationship between compressive strength and constraint expansion strain
At demolding: material age 20 hour.

4. Experiment result and discussion

This experiment investigated the influence of steam curing temperature and pre-curing time on expanding agent for powdered high fluidity concrete blended with blast furnace slag fine powder. As a result, the conclusions of this study are as following.

(1) When expanding material is added for enhancing demolding strength, it was found that the maximum temperature of steam curing is required to be about 55 ° C or more.

(2) The restraint expansion strain due to the addition of the expansion agent increases as the preliminary time increases. However, the effect of enhancing compressive strength was relatively small, and it was almost the same for both type of expansion agent.

(3) When the maximum temperature of steam curing is increased, constraint expansion strain due to the addition of the expansion material is increased in the ettringite-based expansion material, but it decreases in the lime type expansion material.

5. References

[1] Maurice Levitt, "Precast Concrete: Materials, Manufacture, Properties and Usage, Second Edition", CRC Press, September 2007

[2] F.Cassagnabère, G.Escadeillas, and M.Mouret, " Study of the reactivity of cement/metakaolin binders at early age for specific use in steam cured precast concrete", Magazine of Construction and Building Materials, Volume 23, Issue 2, February 2009, pp.775-784

[3] D.W.S. Ho, C.W. Chua, and C.T. Tam, "Steam-cured concrete incorporating mineral admixtures ", Magazine of Cement and Concrete Research, Volume 33, Issue4, April 2003, pp.595-601

[4] J.JBrooks, Megat Johari, "Effect of metakaolin on creep and shrinkage of concrete", Magazine of Cement and Composites, Volume 23, Issue 6 December 2001, pp.495-502

[5] D. Chopin, O. Francy, S. Lebourgeois, and P. Rougeau, "Creep and shrinkage of heat-cured self-compactiong-concrete (SCC)", Book of International RILEM Symposium on Self-Compacting Concrete, August 2003, pp.672-683

[6] Maher K. Tadros, Amin Ghali, Arthur and W. Meyer, "Prestressed Loss and Deflection of Precast Concrete Members", Magazine of PCI Journal, Volume 30, Issue 1 January-February 1985, pp.114-141,

[7] Zdenek P. Bazant, "Prediction of Concrete Creep Effects Using Age-Adjusted Effective Modulus Method", Magazine of International Concrete Abstracts Portal, Volume 69, Issue 4, April 1972, pp.212-219
[8] S. Wild, J. M. Khatib, and L. J. Roose, "Chemical shrinkage and autogenous shrinkage of Portland cement—metakaolin pastes", Magazine of Advances in Cement Research, Volume 10, Issue 3, July 1998, pp.109-119
[9] T.Ramlochan, P.Zacarias, M.D.A.Thomas, and R.D.Hootona, "The effect of pozzolans and slag on the expansion of mortars cured at elevated temperature: Part I: Expansive behavior", Magazine of Cement and Concrete Research, Volume 33, Issue 6, June 2003, pp.807-814
[10] D. W. Hobbs, "Expansion and Cracking in Concrete Associated with Delayed Ettringite Formation", Magazine of International Concrete Abstracts Portal, Volume 177, January 1999, pp.159-182
[11] P.F. McGrath, R.D. Hooton, "Influence of binder composition on chloride penetration resistance of concrete", Magazine of American Concrete Institute, Volume 1, 1997, pp.331-347
[12] Low Sui Pheng, Choong Joo Chuan, "Just-in-Time Management of Precast Concrete Components", Journal of Construction Engineering and Management, Volume 127, Issue 6, December 2001, pp.494-501
[13] Liang, Er-Jun, "Negative Thermal Expansion Materials and Their Applications: A Survey of Recent Patents", Publisher of Bentham Science Publishers, Volume 3, Number 2, June 2010, pp.106-128 (23)

Acknowledgments
The authors want to thank Kazuki Hirabe, Shin-ichiro Hashimoto, Shegeyuki Date. This research was supported by the Tsurumi Concrete Corporation and Tokai University Faculty of Engineering Department of Civil Engineering Department of Construction Materials Laboratory Students.

IOP Conf. Series: Materials Science and Engineering **264** (2017) 012012 doi:10.1088/1757-899X/264/1/012012

Fracture simulation of structural steel at elevated temperature using XFEM technique

Wenyu Cai[1]

1 Assistant Professor, Hefei University of Technology, Hefei, China
E-mail: wenyu.cai@hfut.edu.cn

Abstract. Modeling fracture is important for predicting the actual behavior of structural steel in fire, where fracture of steel members is often the failure mode that affects the overall strength and deformation capacity of steel members. Therefore, in this paper, recent research on modeling steel fracture at elevated temperature is presented. Extended Finite Element (XFEM) together with fracture criterion of principle logarithmic strain is used to simulate the fracture behavior of three different types of tensile test specimens named MA, MB and MC made from American structural steel ASTM A992 at elevated temperature up to 1000°C. The simulation results are then compared with test results to validate the accuracy of simulations. In addition, study is conducted to evaluate how sensitive the predicted engineering stress-strain curves and fracture initiation point on the curves to the selected value of principle logarithmic strain at fracture. Finally, the generalized principle logarithmic strains at each temperature case are proposed for fracture simulation of ASTM A992 steel specimens.

1. Introduction

Fracture behavior of steel at elevated temperature is of great interest because it can significantly influence the response of steel structure in fire. One example of collapse of steel building mainly due to connection fracture is WTC building collapse in US in fire event after severe attacks. It was also observed that steel fracture in tension was a common failure mode during heating or cooling stage in fire events. However, to accurately predict fracture behavior of steel in fire is still difficult due to lack of enough test data and studies. Therefore, in order to propose an approach to predict fracture behavior of steel in fires with reasonable accuracy, standard steel specimens under unaxial tension at room temperature and elevated temperature up to 1000°C with 100°C intervals are investigated and simulated using advanced computational method such as finite element method. The obtained engineering stress-strain curves and failure mechanism of structural steel are compared with several tests data to verify the proposed approach.

1.1. Theoretical background for fracture

In general, fracture can be directly defined as the separations of materials into several pieces as shown in Figure 1. Normally, two categories of fracture modes including brittle fracture and ductile fracture are observed in building materials. Based on several experimental observations, for brittle fracture modes, materials usually undergoes quite little plastic deformation before onset of fracture and cracks will propagate rapidly after fracture initiates; for ductile fracture modes, before the formation of fracture materials developed quite significant plastic deformation and cracks will propagate relatively slowly after onset of fracture. As for structural steel in tension at elevated temperature, it demonstrates ductile fracture modes since it develops quite significant plastic deformation before onset of fracture. Therefore, the research described herein will focus on predicting the ductile fracture behavior of steel.

Ductile fracture initiation in steel can be usually recognized from engineering stress-strain curves as illustrated in Figure 2. Fracture usually initiates once the stiffness of the material begins to degrade significantly.

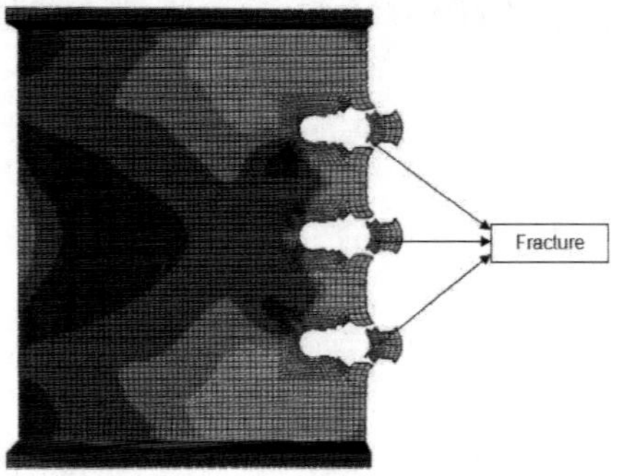

Figure 1. Fracture captured in simulated beam section.

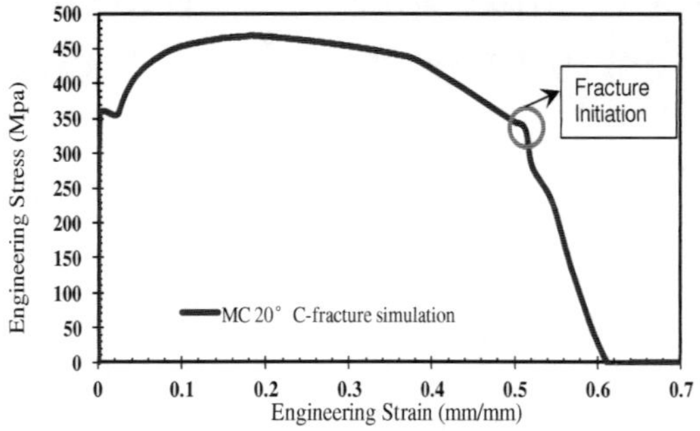

Figure 2. Fracture initiation point captured from engineering stress-strain curve of structural steel.

1.2. Possible approaches to simulate fracture of steel under uniaxial tension

Several fracture initiation models have been proposed in the literatures to simulate fracture behavior of ductile metals. This includes models proposed by McClintock (1968), Rice and Tracey (1969), Hancock and Mackenzie (1976), Johnson and Cook (1985), Hooputra (2004) and Gurson (1977) model [1]-[6] etc. In computational finite element analysis programs, several of these fracture initiation models have been implemented. Fracture simulation of structural steels can be performed using these fracture models together with fracture propagation rules. Nevertheless, accurate fracture model parameters depending on material properties are always required in order to generate accurate prediction results. Calibration of the fracture models parameters still needs lots of efforts and significant numbers of tests and simulations.

Another approach to simulate fracture behavior of ductile metals is using extended finite element method which is designated as XFEM. Fracture models and parameters are not required if XFEM is used. XFEM technique has been used for fracture prediction of ductile metals by several scholars and reported in literatures. Schiavone et al [7] used both Gurson model and XFEM to capture fracture behavior in Aluminium specimens with different notched shapes and observed that the crack shapes were predicted correctly using two approaches, with the exception of the square-notch case using XFEM. Hosseini-Toudeshky and Jamalian [8] performed fracture modeling for bulk Aluminium Al5083 using XFEM and predicted the ductile fracture behavior of Al5083 very well by demonstrating the comparison of stress-strain behavior results between simulations and tests. Vajragupta et al [9] employed XFEM technique to capture fracture behavior of dual phases steel (DP steel), where both ductile fracture and brittle fracture were observed, and demonstrated the capability of XFEM for predicting both ductile and brittle fracture in the same specimen. Ramazani et al [10] investigated the fracture behavior of DP600 steel with various chemical contents and demonstrated that XFEM captured fracture behavior of different types of DP600 steel quite well. It should be noted that the above-mentioned researches using XFEM were all under ambient temperature. According to the knowledge of author, there are very limited studies reported in literatures on fracture simulations of steel at elevated temperatures using XFEM. In this study, XFEM will be investigated and implemented to capture fracture behavior of structural steels at both ambient and elevated temperatures.

1.3. Background of extended finite element method

The extended finite element method (XFEM), although not a new technique for simulation, can be used for fracture simulation for both brittle and ductile failures. XFEM is based on the concept of partition of an element and allows the presence of discontinuities in an element by enriching degrees of freedom with special displacement functions as shown in Equation (1) [11]. In Equation (1), the first term in bracket represents the basic nodal function of all the nodes in element. The second terms considers the node whose shape function is cut by an interior crack, and the third term takes account of nodes whose shape function is cut by crack tips. In XFEM, initial representation of a crack location is not required and arbitrary crack path can be generated between different nodes.

$$u^h(x) = \sum_{I\in N} N_I(x)[u_I + \underbrace{H(x)a_I}_{I\in N_\Gamma} + \underbrace{\textstyle\sum_{\alpha=1}^{4} F_\alpha(x)b_I^\alpha}_{I\in N_\Lambda}] \tag{1}$$

- u_I: Nodal DOF for conventional shape function N_I
- Heaviside enrichment term
  $H(x)$: Heaviside distribution
  a_I: Nodal enriched DOF (jump discontinuity)
  N_Γ: Nodes belonging to elements cut by crack
- Crack tip enrichment term
  $F_\alpha(x)$: Crack tip asymptotic functions
  b_I^α: Nodal DOF (crack tip enrichment)
  N_Λ: Nodes belonging to elements containing crack tip

Two distinct types of fracture modelling can be used in XFEM techniques. The first one is called cohesive segments approach using traction-separation laws based on the maximum principle stress, normal stress, maximum principle strain or normal strain [11]. For this type of fracture simulation, once the principle strain or stress in element reaches critical values defined in material properties, the crack initiates and propagates until the entire member fractures. The other is linear elastic fracture mechanics approach based on the traditional brittle fracture mechanics [11]. In general, cohesive segments approach is applicable for both brittle and ductile fracture, but linear elastic fracture mechanics approach is only applicable for brittle fracture mechanics [11].

Several experimental results suggest that structural steel demonstrates ductile fracture behaviour at elevated temperature. Therefore, XFEM together with cohesive segments approach are employed in this study to predict fracture behaviour of American structural steel ASTM A992 at elevated temperature up to 1000°C. The process of simulation will be elaborated in the following section.

2. Procedure for simulating fracture behaviour of structural steel

2.1. Steel specimens from tests

The tensile test steel specimens made from American structural steel ASTM A992 used for fracture simulations were reported by Lee et al [12]. These were among the limited test of structural steel in tension at elevated temperature providing adequate data from the beginning of load up to fracture. The ASTM A992 materials used for the steel specimens were defined as MA, MB and MC which were cut from the flange and web of wide-flange sections of ASTM A992 steel. The dimensions of three different specimens are shown in Figure 3. The tension tests were conducted as steady-state temperature tests. The obtained engineering stress-strain data of three specimens were recorded from 20°C up to 1000°C.

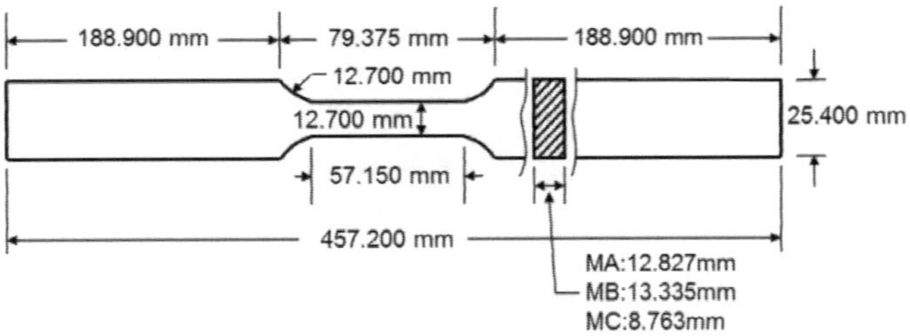

Figure 3. Steel specimens from Lee et al's tests [12].

2.2. Simulation of steel specimens in ABAQUS

Detailed models of above tested specimens were developed in finite element program ABAQUS in version 6.12. In order to obtain the simulation results with reasonable accuracy, material properties of steel should be defined correctly and these include true stress and strain response together with material fracture criterion. It should be noted that true stress represented by loading divided by actual cross sectional area of specimen and true strain represented by instantaneous increment of elongation, cannot be directly measured from tests results due to the change of cross sectional areas in the loading process as well as the complicated stress state after onset of necking. However, accurate true stress and true strain data can be obtained from engineering stress and strain data through calibrating detailed finite element models of tension coupons to match measured engineering stress-strain data.

Once the true stress and strain data are ready for input to material properties, the next part is fracture modelling using XFEM procedure. The detailed specimen model was generated in ABAQUS program with three-dimensional solid brick element (C3D8R). Similar to the boundary conditions in test, one end of the specimen was fixed in each direction, and the other end was free along the axial direction but fixed in other directions. The loading was applied as target displacement with the value of approximately 25mm. The mesh size of the model was defined as approximately 2 mm based on the mesh sensitivity study.

At the beginning of the stage, fracture was not defined in the program; therefore, the simulated engineering stress-strain curves were extended beyond the point at fracture (Figure 4). The aim of this beginning stage is to capture fracture point from engineering stress-strain curves so that the principle logarithmic strain also defined as principle true strain along the loading direction at this point can be calibrated from ABAQUS outputs and further used as the fracture criterion of traction-separation laws. Taking MA specimens at 300°C as example shown in Figure 4, simulation without fracture was compared with test data to locate the fracture point at which the stiffness of material degraded much more significantly than other regions. Then the principle logarithmic strain at this point can be obtained from simulation results and can be taken from the element at the centre of necked area of the model, since fracture initiates at the centre of necked area in specimen according to several experimental observations [13]. Therefore, obtained fracture principle logarithmic strain of necked region was 0.76 mm/mm for MA specimen at 300°C from ABAQUS (see Figure 5).

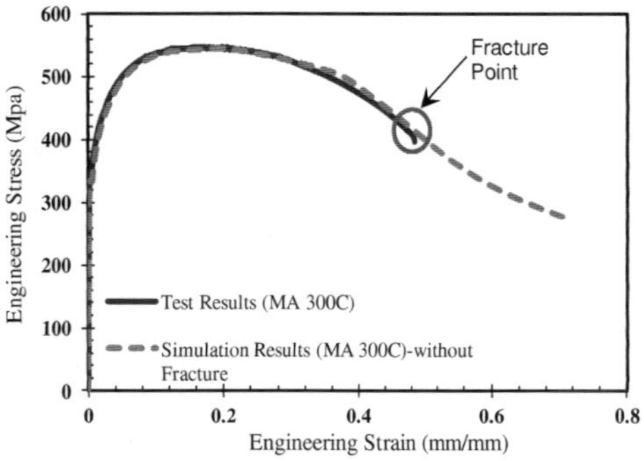

Figure 4. Engineering stress-strain curve of MA at 300°C from test and simulation without fracture.

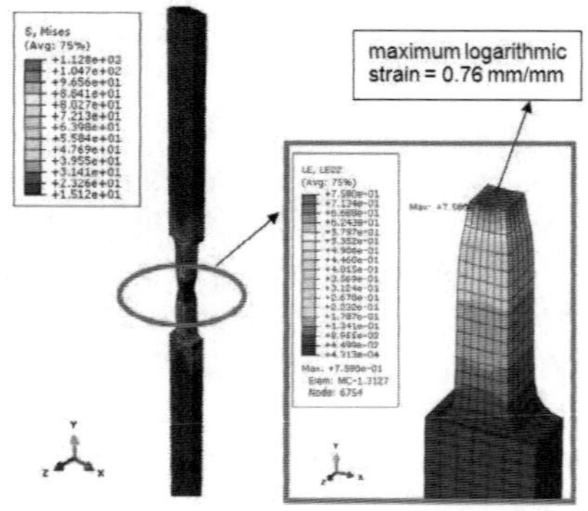

Figure 5. Maximum logarithmic strain at centre of specimen.

Once the principle logarithmic strain at fracture was determined, the fracture simulation can be conducted subsequently. By defining material properties of "Damage for Traction Separation Laws" with maximum principle strain equal to 0.76, the fracture criterion was defined. The displacement at fracture can be determined using principle logarithmic strain multiplied by the mesh size in the necked area so that the mesh size effects on the results can be considered. For ductile fracture simulation, cohesive segment approach in XFEM would be appropriate; therefore, cohesive behaviour of the segment should be assigned in the interaction properties to capture material separation in the model. Possible fracture location should be assigned in the model by creating crack domain. The entire model can be assigned with crack domain so that any crack formation in the model can be captured. Further, in order to make the crack formation and propagation visible, in additional to the force, stress and strain output, failure state of "PHILSM" representing fracture surface and "STATUSXFEM" indicating state of cracked elements should be selected as outputs as well.

The formation and development of fracture are shown in Figure 6. The crack was first formed in the necked area in the middle of specimens and was further propagated throughout the cross section. Finally, the specimen was cut by the formed fracture surface and divided into two separate parts. The same phenomenon can be also observed in tests.

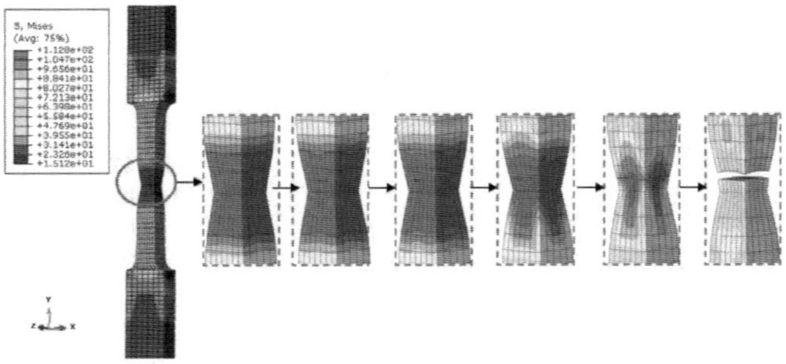

Figure 6. Fracture formation process in simulation.

The simulated engineering stress-strain curves were compared with tested curves as shown in Figure 7. The failure modes from simulation were compared with those from tests as illustrated in Figure 8. The results for MA specimen at 300°C from simulation agree with tests results quite well. The comparison suggests that the XFEM approach can capture the fracture behaviour of simulated specimen quite well.

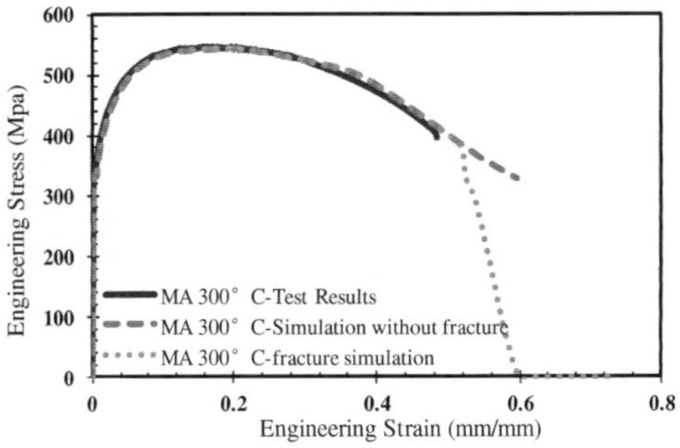

Figure 7. Comparison of test and simulation results of engineering stress-strain curves for MA 300°C.

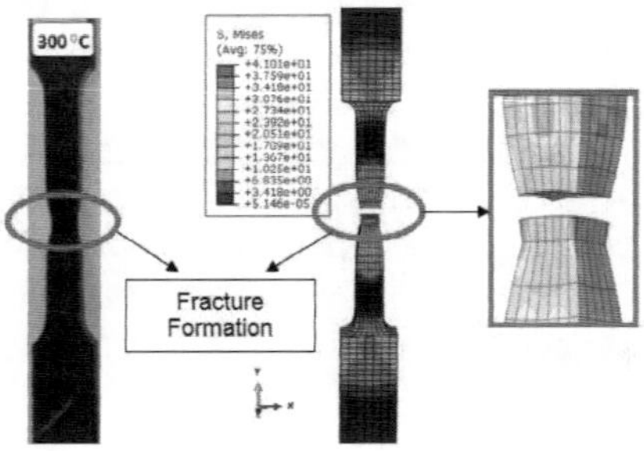

Figure 8. Failure mode of MA specimen at 300°C from test and simulation.

3. Observations from simulation results

3.1. Comparison of simulation and test results at different temperature cases

In this research, three different types of ASTM A992 steel specimens MA, MB and MC were investigated under uniaxial tension and at elevated temperatures. The obtained results of each specimen for each temperature case were illustrated in Figure 9. These figures demonstrate the comparison of engineering stress-strain curves between tests and simulations. Some difference is still observed in the comparison, especially at higher temperatures such as 700°C or 800°C, however, in general, the simulated results match tested results quite well indicating the practicability of XFEM approach to simulating fracture behaviour of steel at elevated temperature.

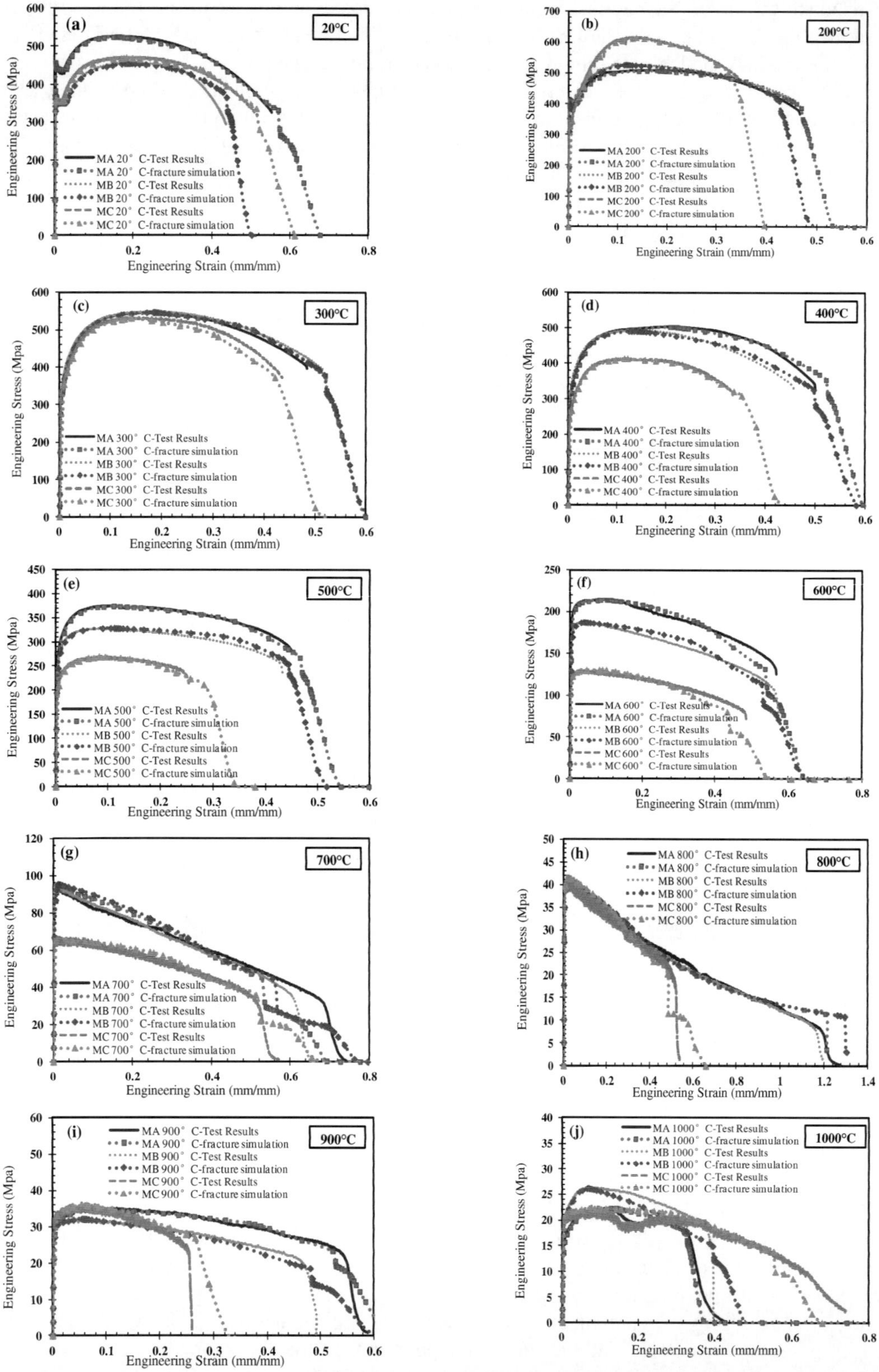

Figure 9. Comparison of engineering stress-strain curves of three specimens between tests and simulations for each temperature case.

3.2. Sensitivity study of fracture simulation results to selected principle logarithmic strain

In this research, the principle logarithmic strain at fracture of the central element in the necked area was used as critical strain to initiate the fracture. The sensitivity of simulation results to the selected principle logarithmic strains is of interest since the sensitivity study will be useful for generalizing the principle logarithmic strain at fracture and predicting the fracture of other ASTM A992 steel specimens besides MA, MB and MC.

Taking MA specimen at 200°C as example, the sensitivity study of the fracture principle logarithmic strain is shown in Figure 10(a) and Table 1. The simulation was conducted for principle logarithmic strain equal to 0.66 mm/mm, 0.76 mm/mm and 0.86 mm/mm, which represents 15% to 30% variation of principle logarithmic strain for the base case of 0.66 mm/mm strain. From Figure 10(a), change of fracture principle logarithmic strain within this range has little effect on the predicted maximum engineering stress. However, results suggest that there is some variation in the predicted engineering stress and strain at fracture as fracture strain is varied.

The same sensitivity study was also performed for MB specimen. Taking MB specimen at 400°C as example, with the change of fracture principle logarithmic strain from 13% to 25%, the maximum predicted engineering stress demonstrates quite little change (See Figure 10(b)), but there is some variation in the predicted engineering stress and strain at fracture (see Figure 10(b) and Table 1).

As for sensitivity study for MC specimen at 500°C, with the variation of fracture principle logarithmic strain from 19% to 38%, the predicted maximum engineering stress is not varied, however, the predicted engineering stress and strain at fracture are changed with the variation of principle logarithmic strain (see Figure 10(c) and Table 1).

For MA, MB and MC specimen at other temperature cases, the similar observations can be reached when the selected principle logarithmic strain at fracture was varied within approximately 15% to 30%. Based on this limited sensitivity study, it is observed that prediction of fracture based on the principle logarithmic strain criterion is only mildly sensitive to small variation of selected logarithmic strain.

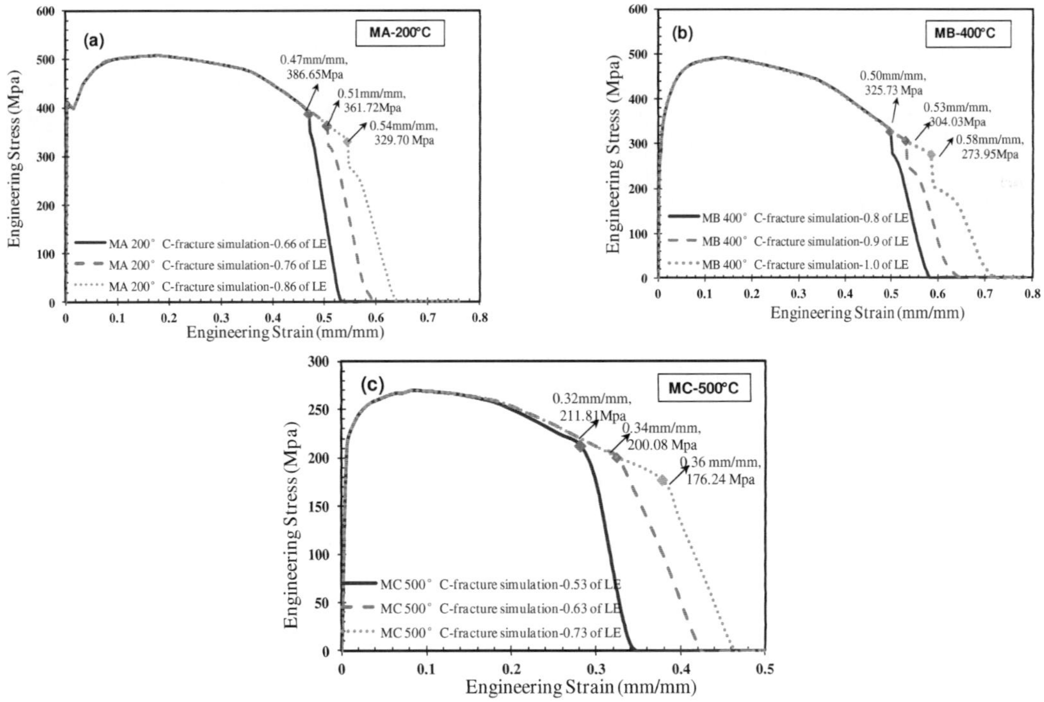

Figure 10. Sensitivity study for selected fracture strain.

Table 1. Example of sensitivity study of fracture strain.

Specimens	Principle logarithmic strain at fracture	**Variation** of principle logarithmic strain **in percentage**	Predicted engineering strain at fracture	**Variation** of predicted engineering strain at fracture **in percentage**
MA at 200°C	0.66	0%	0.47	0%
	0.76	15%	0.51	8%
	0.86	30%	0.54	15%
MB at 400°C	0.80	0%	0.50	0%
	0.90	13%	0.53	6%
	1.00	25%	0.58	15%
MC at 500°C	0.53	0%	0.32	0%
	0.63	19%	0.34	6%
	0.73	38%	0.36	13%

3.3 Generalized fracture strains for fracture simulation of structural steel at elevated temperature

According to the simulation results of MA, MB and MC specimens at room temperature and elevated temperature, the simulation results agree with tests results quite well in general. The principle logarithmic strain criterion in simulations is summarized and generalized as Figure 11. Below and inclusive 600°C, the fracture principle logarithmic strains are all lower than 1.0 mm/mm, but for 700°C and 800°C temperature cases, they are higher than 1.0 mm/mm, and then as for 900°C and 1000°C cases, they are dropped below 1.0 mm/mm again.
From Table 2, the critical logarithmic strains in simulations for MA, MB and MC were relatively consistent from 200°C to 600°C as well as 1000°C. The maximum difference of the fracture principle logarithmic strains between that for each specimen and average value for three specimens are no more than 8%. According to previous sensitivity study for selected principle logarithmic strain, the variation of the strain within approximately 15%, the variation of predicted results were approximately 6% to 8%. Therefore, using average value of critical strains may still predict the fracture behavior of those specimens with reasonable accuracy. But cautions should be made when average values of principle logarithmic strain at 20°C, 700°C, 800°C and 1000°C are used for fracture simulations, because large difference exists in principle logarithmic strains among MA, MB and MC indicative of inconsistence in fracture criterion for simulations at these temperature cases by using XFEM. More tests on different types of specimens are still needed at these temperatures to investigate this inconsistence.

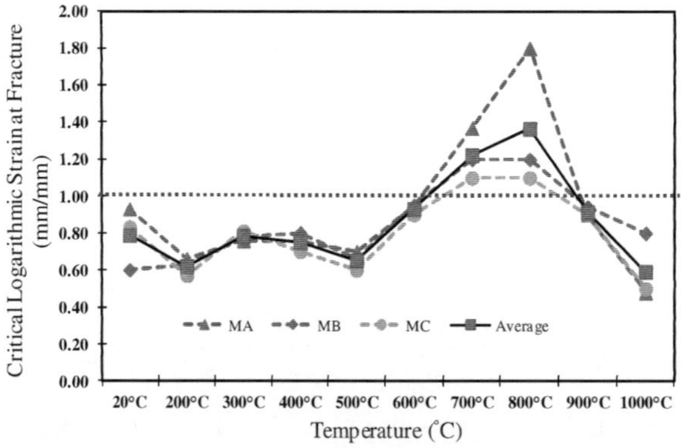

Figure 11. Summary of critical logarithmic strain at fracture for different temperature cases.

Table 2. Summary of principle logarithmic strain at fracture for fracture simulation of ASTM A992 steel at elevated temperature.

Temperature	Principle Logarithmic Strain at Fracture			
	MA	MB	MC	Average
20°C	0.93	0.60	0.83	**0.79**
200°C	0.66	0.63	0.57	**0.62**
300°C	0.76	0.78	0.81	**0.78**
400°C	0.75	0.80	0.70	**0.75**
500°C	0.70	0.66	0.60	**0.65**
600°C	0.94	0.95	0.90	**0.93**
700°C	1.37	1.20	1.10	**1.22**
800°C	1.80	1.20	1.10	**1.37**
900°C	0.90	0.94	0.90	**0.91**
1000°C	0.48	0.80	0.50	**0.59**

4. Conclusion and future research

Work presented in this study has demonstrated the feasibility of simulating fracture behaviour of ASTM A992 steel tension specimens at elevated temperatures using extended finite element method (XFEM). In addition, fracture principle logarithmic strain as fracture criterion for XFEM approach was also generalized for fracture simulations for structural steels. Several conclusions are reached as follows:

- Fracture behaviour of steel tension specimen can be simulated not only by using traditional fracture models but also by using extended finite element method. In general, the principle logarithmic strain or principle stress is needed to initiate the fracture in simulations.
- Based on study of different type of ASTM A992 steel specimens, the simulation results using XFEM approaches match test results quite well at the elevated temperatures above 200°C and below 1000°C.
- As for the fracture criterion described in this research, study suggests that the fracture principle logarithmic strains of three different specimens at same temperature case are quite consistent in general; prediction of fracture based on the principle logarithmic strain criterion is only mildly sensitive to small variation of selected logarithmic strain; and the maximum predicted engineering stress is not affected by the variation of fracture logarithmic strains if other material properties are not changed.

In general, this study has demonstrated the potential for reasonable computational prediction of fracture for ASTM A992 tensile test specimens under ambient and elevated temperatures. The proposed modelling procedure and generalized logarithmic fracture strain can be used as reference for fracture simulation of the same material. However, there are still some limitations existing in this study. Firstly, the work is mainly based on the limited experimental data from three different types of steel specimens made from ASTM A992; therefore, further research work is also needed on a sufficient number of samples of steel to quantify variability and uncertainty in fracture prediction. In addition, large difference observed in logarithmic strains at fracture at same temperature may be due to either the material variability or effects from modelling technique such as mesh size, element type, model geometry etc. Thus, modelling technique effects on fracture simulation accuracy should be investigated in the further research.

5. References

[1] McClintock A Frank 1968 A criterion for ductile fracture by the growth of holes *J Appl. Mech.* **35**(2) 363-371

[2] Rice J R and Tracey D M 1969 On the ductile enlargement of voids in triaxial stress fields *J. Mech. Phys. Solids.* **17** 201-217

[3] Hancock J W and Mackenzie A C 1976 On the mechanisms of ductile failure in high-strength steels subjected to multi-axial stress-states *J. Mech. Phys. Solids.* **24** 147-169

[4] Johnson R Gordon and Cook H William 1985 Fracture characteristics of three metals subjected to various strains, strain rates, temperatures and pressures *Eng. Fract. Mech.* **21**(1) 31-48

[5] Hooputra H, Gese H, Dell H and Werner H 2004 A comprehensive failure model for crashworthiness simulation of aluminium extrusions *Int. J. Crashworthiness.* **9**(5) 449-464

[6] Gurson L Arthur 1977 Continuum theory of ductile rupture by void nucleation and growth: Part I—Yield criteria and flow rules for porous ductile media *J. Eng. Mater. Technol.* **99**(1) 2-15

[7] Schiavone A, Abeygunawardana-Arachchige G and Silberschmidt V V 2016 Crack initiation and propagation in ductile specimens with notches: experimental and numerical study *Acta. Mechanica.* **227**(1) 203

[8] Hosseini-Toudeshky H and Jamalian M 2015 Simulation of micromechanical damage to obtain mechanical properties of bimodal Al using XFEM *Mech. Mater.* **89** 229-240

[9] Vajragupta N, Uthaisangsuk V, Schmaling B, Münstermann S, Hartmaier A and Bleck W 2012. A micromechanical damage simulation of dual phase steels using XFEM *Comput. Mater. Sci.* **54** 271-279

[10] Ramazani A, Abbasi M, Kazemiabnavi S, Schmauder S, Larson R and Prahl U 2016 Development and application of a microstructure-based approach to characterize and model failure initiation in DP steels using XFEM *Mater. Sci. Eng., A* **660** 181-194

[11] Hibbitt, Karlsson and Sorensen Inc 2012 *ABAQUS Version 6.12 User's Manual.*

[12] Lee J, Morovat M A , Hu G, Engelhardt M D and Taleff E M, 2013. Experimental investigation of mechanical properties of ASTM A992 steel at elevated temperatures *Eng. J. AISC.* **50**(4) 249-272

[13] Kanvinde M Amit 2004 Micromechanical simulation of earthquake-induced fracture in steel structures *Ph.D. dissertation,* Stanford University. Civil & Environmental Engineering Department

Acknowledgments

The Author wishes to thank Professor Michael D. Engelhardt from The University of Texas at Austin for providing valuable test data as reference. This work was supported in part by a grant from Young Scholar Research Funding from Hefei University of Technology in China.

IOP Conf. Series: Materials Science and Engineering **264** (2017) 012013 doi:10.1088/1757-899X/264/1/012013

Applications of graphite-enabled phase change material composites to improve thermal performance of cementitious materials

Mingli Li[1, a], Zhibin Lin[1, b], Lili Wu[2, c], Jinhui Wang [3, d] and Na Gong [3, e]

[1]Department of Civil and Environmental Engineering, North Dakota State University, Dept of CE Room 203, 1410, North 14th Avenue, Fargo, ND58105, USA
[2]School of Mechanics and Civil Engineering, China University of Mining and Technology, Beijing, 100083, China
[3]Department of Electrical and Computer Engineering, North Dakota State University, Fargo, ND58105, USA

[a]mingli.li@ndsu.edu, [b]Corresponding author: zhibin.lin@ndsu.edu, [c]jennywll@163.com, [d]jinhui.wang.1@ndsu.edu, [e]na.gong@ndsu.edu

Abstract. Enhancing the thermal efficiency to decrease the energy consumption of structures has been the topic of much research. In this study, a graphite-enabled microencapsulated phase change material (GE-MEPCM) was used in the production of a novel thermal energy storage engineered cementitious composite feathering high heat storage capacity and enhanced thermal conductivity. The surface morphology and particle size of the microencapsulated phase change material (MEPCM) were investigated by scanning electron microscopy (SEM). Thermal properties of MEPCM was determined using differential scanning calorimetry (DSC). In addition, thermal and mechanical properties of the cementitious mortar with different admixtures were explored and compared with those of a cementitious composite. It was shown that the latent heat of MEPCM was 162 J/g, offering much better thermal energy storage capacity to the cementitious composite. However, MEPCM was found to decrease the thermal conductivity of the composite, which can be effectively solved by adding natural graphite (NG). Moreover, the incorporation of MEPCM has a certain decrease in the compressive strength, mainly due to the weak interfaces between MEPCM and cement matrix.

1. Introduction

In recent years, building structures consume enormous amounts of energy due to population growth, the increasing demand for thermal comfort and the fact that more and more people spend their time indoor [1-3]. Decreasing the energy consumption of structures by enhancing their energy efficiency has become significant research topic worldwide during the last two decades [4, 5]. Incorporation of thermal energy storage systems in buildings is regarded as a feasible way of enhancing energy efficiency [6]. Compared to the other two thermal energy storage types, which are sensible heat storage and chemical reaction heat storage, latent heat storage with PCMs has remarkable advantages like high-energy storage density and the isothermal nature during the energy storage process, rendering it an attractive material for saving energy and improving occupant comfort of buildings [7-9].

Cementitious materials are considered suitable for incorporation of PCMs because it is one of the most extensively used construction material around the world. The large thermal mass of concrete walls, ceilings, and floors with PCMs can be well used to store energy during the day time and release it during night time. To effectively enhance the thermal storage capacity of concrete, much research [10-12] has been conducted by using phase change materials (PCMs) during recent decades. Xu et al. [3] presented their study on incorporating a paraffin/diatomite composite phase change material to concrete for improving the thermal performance. Their results revealed that the PCMs effectively enhance the ductile performance and heat energy storage capacity. Sharifi and Sakulich [13] stated that adding PCM in mortar increases the thermal mass and delays temperature changes. However, a compressive strength loss was found in both researches. It is caused by significant disparity between the cementitious material and the microencapsules. In addition, partially leakage caused by the damage of the microencapsules during the mixing process and subsequent interference with the surrounding materials is another reason resulting in the compressive strength loss.

Among different types of PCMs, microencapsulated PCM (MEPCM) turns out a promising candidate due to its non-leakage property and relatively high heat of fusion. It is composed of n-octacosane as core PCM and a polymer shell for keeping liquid PCM from leaking during the phase change process. However, both the PCM inside and the polymer shell have an even lower thermal conductivity than cement matrix, so the incorporation will cause temperature gradient when a temperature difference exists, further resulting in thermal cracks of the cementitious composite. Several studies have been directed to the improvement of concrete thermal conductivity [14-16]. Lie et al. [14] investigated the use of steel fiber in the concrete mixture to improve thermal properties. Fu et al. [16] reported their tests on the use of different additives, including silica fume, latex, methylcellulose and carbon fibers, on the thermal conductivity of cement paste.

Clearly, the literature review shows that although there are studies that addressed either of two aspects (i.e., the low thermal conductivity and low thermal storage capacity), very limited information is available to consider both in the design of the building materials. This study is to investigate the feasibility of using graphite-enabled PCM composite for the enhanced thermal management of cementitious material. A series of laboratory tests such as SEM and DSC were performed to characterize the morphology and thermal behavior of different mixtures. In addition, heat transfer and compressive strength test were conducted to examine the thermal and mechanical properties of the composite.

2. Experimental program

2.1 Specimen Preparation

Technical-grade MEPCM purchased from Microtek Laboratories, Inc. were used as the PCMs. They are very small bi-component particles consisting of a core material— octadecane – and an outer shell or capsule wall. The melting point of MEPCM is 109.4 °F (43°C). NG (Grade 3772) donated by Asbury Carbon, Inc. was selected as a heat exchanger to enhance the thermal conductivity of the composite, which has an average particle size of 300 μm. In order to reduce the number of influence factors and focus the research on the thermal property, conventional mortar consisting of water, Type I Portland cement, and natural sand was prepared, excluding the effect of coarse aggregate. To achieve more homogeneous mortar mixtures, viscosity modifying admixture (VMA) and water reducer were used at a weight ratio of 7.95:1000 to cement. In this test, seven mix proportions were used with the same water/cement ratio of 0.35 (Table 1). The specimen of heat transfer test and compressive strength test was prepared by casting a 50×50×50 mm3 (2×2×2 in3) cuboid mold. The samples were compacted by tempers in two layers.

Table 1. Mix proportion of mortar with different admixtures

Groups	Cement (kg/m3)	Water (kg/m3)	Natural Sand (kg/m3)	VMA (kg/m3)	Water Reducer (kg/m3)	NG (m3/m3)	MEPCM (m3/m3)
Reference	1000	350	857	7.95	7.95	0	0
10% NG	900	315	771.4	7.16	7.16	0.1	0
20% NG	800	280	685.71	6.36	6.36	0.2	0
15% MEPCM	850	297.5	728.57	6.76	6.76	0	0.15
30% MEPCM	700	245	600	5.57	5.57	0	0.3
10%NG +15%	750	262.5	642.86	5.96	5.96	0.1	0.15
20%NG +15%	650	227.5	557.14	5.17	5.17	0.2	0.15

2.2 Sample characterization
The morphology of MEPCM and its interface with cement matrix was explored using a scanning electron microscope (SEM). DSC measurements of MEPCM were carried out on a Differential scanning calorimeter (DSC2910, TA Instrument Inc., USA) under N2 atmosphere. The typical mass of the sample was about 4mg, and the scanning rate was 5°C/min.

2.3 Heat Transfer and Compressive Strength Tests
In the thermocouple heat transfer test, the operating principle is based on the association of a heating device with temperature sensors. A data logger was connected to the thermocouples (type T) to record the temperature increase on the top surface of the specimen during a predetermined temperature range. The hot plate was used as a heating device to provide a constant temperature (80 °C) for the bottom of the specimen.

ASTM C109/109M was followed for the compressive strength test of the mortars. After mixing, mortar was placed in 50mm (2 in) cube plastic molds for 24h. Then they were demolded and cured at room temperature (22 °C) with relative humidity of 16% till the day of testing (3, 7, and 28 days). For each mix, the compressive strength is calculated as the average value of three repeated samples.

3. Results and discussions

3.1 Morphology and Thermal Energy Storage Capacity Investigation
Surface morphology of obtained MEPCM was studied by SEM. Figure 1(a) shows that the particles size of MEPCM is 5-15 μm. During phase change process, the capsule wall can effectively maintain liquid materials and avoid the PCM from leaking. Figure 1(b) indicates that the heat of fusion of MEPCM is 162.0 J/g, which make it a promising material for thermal energy storage application. In addition, the material starts to melt at 41.67 ℃, which is very close to the nominal melting point of 43℃.

(a) SEM image of MEPCM [17]

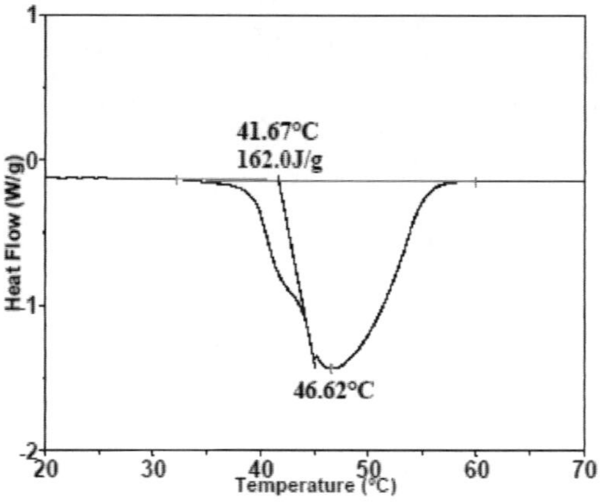

(b) DSC curve of MEPCM

Figure 1. Characterization of MEPCM

3.2 Heat Transfer Speed

Heat transfer test aims at investigating the effect of MEPCM and NG on the thermal conductivity of mortar paste. The heat transfer speed was calculated as the average temperature of three samples cured for 28 days. Figure 2(a) shows that the addition of MEPCM results in an obvious reduction of thermal conductivity. This can be explained by the enhanced air content and the low thermal conductivity of MEPCM. Moreover, the temperature increasing tendency of the sample with 30% MEPCM indicates that phase transition from solid to liquid lasts from 64.7min (38.75°C) to 108.5 min (45°C), which centers on the melting point of MEPCM. During this period, a large amount of latent heat was stored in the composite, which can be released when the ambient temperature drops. Figure 2(b) reveals that NG significantly increases the heat transfer speed of the mortars, rendering it an attractive candidate as a heat exchanger.

The effect of a combination of MEPCM and NG on thermal conductivity of the mortar was presented in Figure 3. It can be observed that NG contributed a lot to the heat transfer speed enhancement of the composite with 15% MEPCM. When 20% NG is added, the heat transfer rate of the composite with 15% MEPCM is even higher than the reference mortar. The addition of a combination of MEPCM and NG can introduce the mortar thermal energy storage capacity without decreasing the thermal conductivity of the cementitious composite.

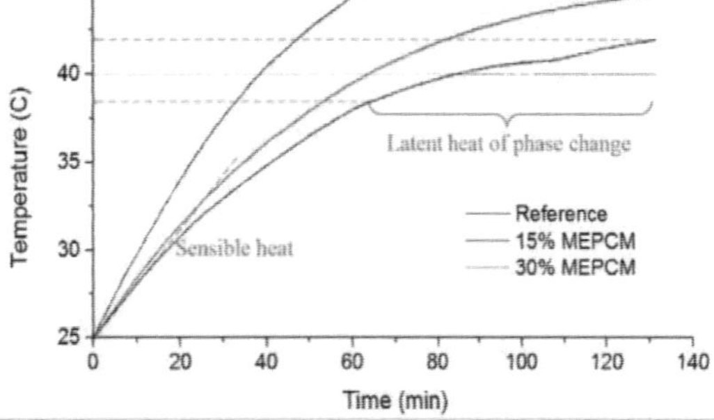

(a) Heat transfer of mortar with MEPCM

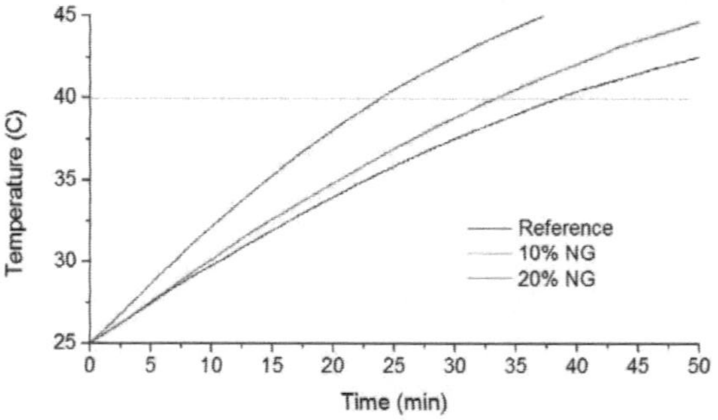

(b) Heat transfer of mortar with NG

Figure 2. Heat transfer speed comparison of mortars with different additives

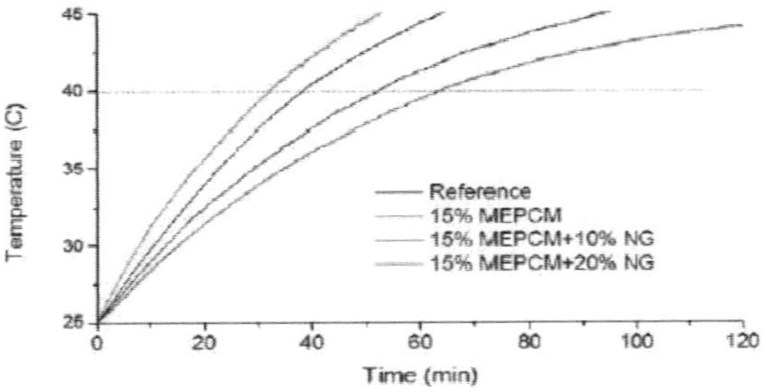

Figure 3. Heat transfer of mortars with a combination of MEPCM and NG

3.3 Compressive strength

The Compressive strength of the mortars was measured in accordance with ASTM C109/C109M. Figure 4(a) shows the compressive strength development of the mortars (cured for 28 days) with different amount of MEPCM and NG. It can be observed that the corporation of different admixtures reduces the compressive strength of mortar in different levels. The reduction effect on compressive strength can be listed as follows: NG together with PCM > NG > MEPCM. This can be explained by the hydrophobic property of the graphite product and MEPCM, which results in a bonding strength reduction of the cement paste. It can be found in Figure 4(b) that there are some gaps between the cement matrix and NG (position 1) as well as MEPCM (position 2), which can result in the compressive strength loss of the composite. Moreover, some PCM capsules can be damaged (see position 3 in Figure 4(b)) with the PCM inside filling surrounding cement matrix during the mixing process, resulting in further compressive strength loss of the cementitious composite.

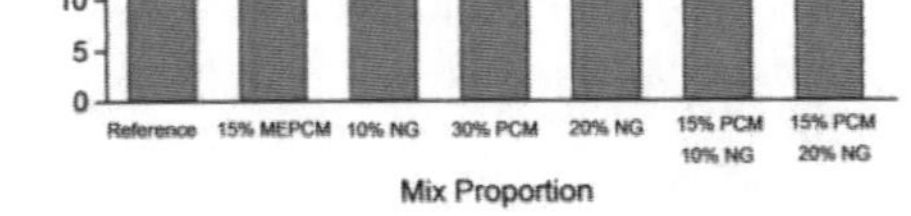

(a) Compressive strength of different mixtures

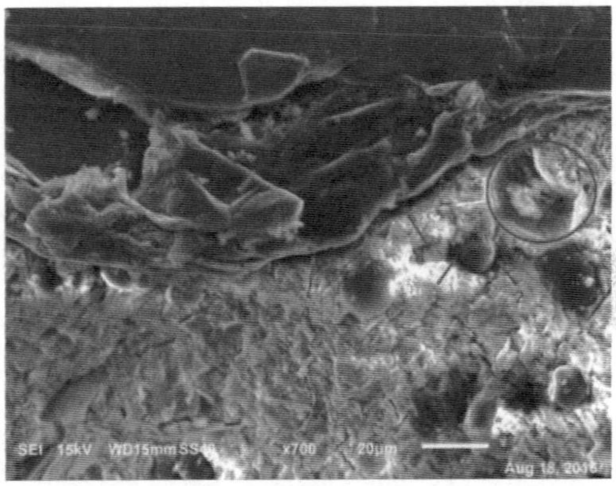

(b) SEM image of mortar with 10% NG and 15% MEPCM

Figure 4. Compressive strength of mortars with different additives and the SEM image of one composite

4. Conclusions

This study developed an innovative graphite-enabled phase change material composite that is capable of inducing thermal energy storage capacity and high thermal conductivity to the cementitious composite. The test results demonstrated that MEPCM in the composite can absorb and store a large amount of latent heat during phase transition from solid to liquid. Moreover, the incorporation of NG significantly increased the thermal conductivity of the mortars. The thermal efficiency of cementitious composites was improved a lot by adding both of MEPCM and NG, without decreasing their compressive strength too much. Further work needs to be carried out to enhance the hydrophilic property of the graphite by surfactant treatment to increase the bond strength of the cement paste and admixtures.

5. References

[1] L. Pérez-Lombard, J. Ortiz, and C. Pout, *A review on buildings energy consumption information* (Energy and buildings, 2008), pp. 394–398.

[2] Z. Ma and S. Wang, *Building energy research in Hong Kong: a review* (Renewable and Sustainable Energy Reviews, 2009), pp. 1870–1883.

[3] B. Xu and Z. Li, *Performance of novel thermal energy storage engineered cementitious composites incorporating a paraffin/diatomite composite phase change material* (Applied Energy, 2014), pp. 114–122.

[4] A. M. Papadopoulos, T. G. Theodosiou, and K. D. Karatzas, *Feasibility of energy saving renovation measures in urban buildings: The impact of energy prices and the acceptable pay back time criterion* (Energy and buildings, 2002), pp. 455–466.

[5] J. Laustsen, *Energy efficiency requirements in building codes, energy efficiency policies for new buildings* (International Energy Agency (IEA), 2008), pp. 477–488.

[6] C. Barreneche, M. E. Navarro, A. I. Fernández, and L. F. Cabeza, *Improvement of the thermal inertia of building materials incorporating PCM. Evaluation in the macroscale.* (Applied energy, 2013), pp. 428–432.

[7] T. C. Ling and C. S. Poon, *Use of phase change materials for thermal energy storage in concrete: an overview* (Construction and Building Materials, 2013), pp. 55–62.

[8] B.Zalba, J. M. Marín, L. F. Cabeza, and H. Mehling, *Review on thermal energy storage with phase change: materials, heat transfer analysis and applications* (Applied thermal engineering, 2003), pp. 251–283.

[9] A. Sharma, V. V. Tyagi, C. R. Chen, and D. Buddhi, *Review on thermal energy storage with phase change materials and applications* (Renewable and Sustainable energy reviews, 2009), pp. 318–345.

[10] A. M. Khudhair and M. F. Mohammed, *A review on energy conservation in building applications with thermal storage by latent heat using phase change materials* (Energy conversion and management, 2004), pp. 263–275.

[11] F. Agyenim, N. Hewitt, P. Eames, and M. Smyth, *A review of materials, heat transfer and phase change problem formulation for latent heat thermal energy storage systems (LHTESS)* (Renewable and Sustainable energy reviews, 2010), pp. 615–628.

[12] A. G. Entrop, H. J. H. Brouwers, and A. H. M. E. Reinders, *Experimental research on the use of micro- encapsulated phase change materials to store solar energy in concrete floors and to save energy in Dutch houses* (Solar Energy, 2011), pp. 1007–1020.

[13] N. P. Sharifi and A. Sakulich, *Application of phase change materials to improve the thermal performance of cementitious material* (Energy and Buildings, 1996), pp. 83–95.

[14] T. T. Lie and V. K. R. Kodur, *Thermal and Mechanical Properties of Steel-fibre-reinforced Concrete at Elevated Temperatures* (Canadian Journal of Civil Engineering, 1996), pp. 511–517.

[15] D. D. L. Chung, *Carbon materials for structural self-sensing, electromagnetic shielding and thermal interfacing* (Carbon, 2012), pp. 3342–3353.

[16] X. L. Fu and D. D. L. Chung, *Effects of silica fume, latex, methylcellulose, and carbon fibers on the thermal conductivity and specific heat of cement paste* (Cement and concrete research, 1997), pp. 1799–1804.

[17] M. L. Li, N. Gong, J. H. Wang, and Z. B. Lin, *Phase Change Material for Thermal Management in 3D Integrated Circuits Packaging* (International Symposium on Microelectronics, 2015), pp. 649-653.

Acknowledgements

The authors gratefully acknowledge the material denotation by Asbury Carbon, Inc. The results, discussion, and opinions reflected in this paper are those of the authors only and do not necessarily represent those of the sponsor.

IOP Conf. Series: Materials Science and Engineering **264** (2017) 012014 doi:10.1088/1757-899X/264/1/012014

Urban renewal based wind environment at pedestrian level in high-density and high-rise urban areas in Sai Ying Pun, Hong Kong

J W YAO[1], J Y ZHENG[1], Y ZHAO[1], Y H SHAO[1] and F YUAN[1*]

1 Built Environment Technology Center, College of Architecture and Urban Planning, Tongji University, Shanghai, China
E-mail: philipyuan007@tongji.edu.cn

Abstract. In high-density and high-rise urban areas, pedestrian level winds contribute to improve comfort, safety and diffusion of heat in urban areas. Outdoor wind study is extremely vital and a prerequisite in high-density cities considering that the immediate pedestrian level wind environment is fundamentally impacted by the presence of a series of high-rise buildings. In this paper, the research site of Sai Ying Pun in Hong Kong will be analysed in terms of geography, climate and urban morphology, while the surrounding natural ventilation has also been simulated by the wind tunnel experiment Computational Fluid Dynamics (CFD). It has found that, the existing problems in this district are the contradiction between planning control and commercial interests, which means some areas around tall buildings are not benefit to the residents because of the unhealthy wind environment. Therefore, some recommendation of urban renewal strategy has been provided.

1. Introduction

Urban renewal plays important role in urban economic development and urban construction, which is also an unavoidable problem in contemporary urban development in developing countries [1]. Despite Hong Kong has only approximately 150 years history, the severe conflict between human and land, leading to urban development strategy with the characteristic of high-density and high-rise, increases the work of urban renewal in Hong Kong. However, based on the urban renewal in Hong Kong, the environmental problems are emerging and may affect the residents' health and comfort. For example, many people died due to severe acute respiratory syndrome (SARS) in Hong Kong in 2003, which led the government to establish a Government Team Clean Committee to improve local urban design policies. In 2006, all development projects funded by the government had to apply a new system of air ventilation assessment (AVA) [2]. Using AVA, urban designers and architects can optimize the natural air ventilation in high-density areas of Hong Kong to avoid stagnant and slow air movement to obtain an appropriate air change rate in urban open public spaces. This is intended to generate a healthy and comfortable outdoor environment. According to previous research focused on Hong Kong, improving air ventilation to create a better wind environment has been demonstrated as the most important consideration in designing and developing sustainable cities, especially under the situation of urban renewal [3].

Sai Ying Pun is situated in the Western District of Hong Kong Island and is a key administrative region for both the Western and Central districts. In the local tongue, Sai means "west" while Ying Pun translates as "camp", generally a military camp. The British military were based in this area and the locality was one of the first developmental zones in wider Hong Kong. Although the existing buildings are in poor repair, most of these are residential buildings and a lot of shops, selling sea food products and traditional Chinese medicinal materials, exist on the ground level (figure 1). In Sai Ying Pun, there are many small lanes and the majority of these are only open to pedestrians. While these lanes may have been quite significant in the past as they provided access to local residential properties, they now primarily function as back alleys and shortcuts between housing units. Furthermore, the area has a primarily aging population and as many elderly people living in this district prefer to socialise outdoors, the stipulation for a safe, comfortable and healthy outdoor wind environment at the pedestrian level is crucial, not only for the inhabitants, but also for those working here. Additionally, there are many similar narrow streets surrounded by tall buildings in other regions of Hong Kong, which may account for Sai Ying Pun being a popular outdoor thermal comfort research case study for researchers. Although the geometry, urban and building layout in this site are applied as a convenient case study with a high-density tall building district to compare the results between wind tunnel tests and CFD simulation, understanding of some aspects of social environment, neighbourhoods, residents' requirement etc. is conducive to analysing and generating some recommendation of urban renewal strategy in Hong Kong and other developing regions facing urban renewal.

Figure 1. Introduction of Sai Ying Pun: (a) location in Hong Kong (blue point) and (b) street view.

The red dashes illustrate the target research site area for wind tunnel testing and CFD simulation (figure 2a and 2b). Connaught Road West and the ocean are north of this site, whilst the Sai Ying Pun Metro Station is on the south. There are approximately 36 tall buildings and podiums that link a number of these. The local meteorological data is always monitored by the nearest meteorological station, and some are in airports (rural area) while some are located at the top of buildings or mountains (urban area). The nearest station around Sai Yin Pun belongs to the latter one, and it is located on the roof of one tall building (approximately 60m height) in the University of Hong Kong, and their straight-line distance is about 300m.

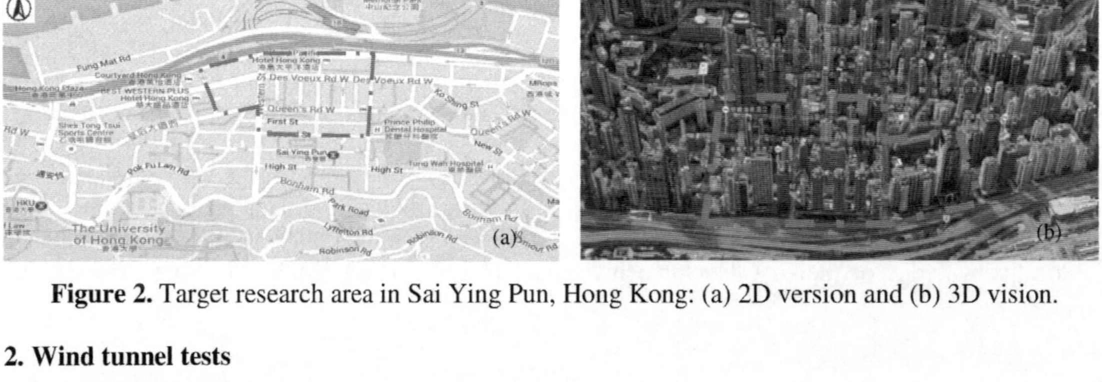

Figure 2. Target research area in Sai Ying Pun, Hong Kong: (a) 2D version and (b) 3D vision.

2. Wind tunnel tests

This section details the equipment, experiment set-up, procedure and also results of wind tunnel tests applied in the outdoor wind environment simulation in Sai Ying Pun to modify the parameter settings of CFD simulation. Wind tunnel tests are carried out to evaluate the interference effects attributed to several medium-rise and high-rise buildings.

2.1. Test cases and experiment set-up

The experiments were performed in the boundary layer wind tunnel at the Department of Civil Engineering, University of Hong Kong. The working section is 12m long, 3m wide and 1.8m high. To simulate natural wind conditions, 8m long fetch of floor roughness elements have been applied (figure 3b). The inlet wind for simulation is provided by the fan from behind the wooden rail fence. Opposite the simulation room (figure 3a), the target building models could be fixed on a rotatable round plate (2.5m diameter). After passing through the building target site, the air will enter another empty room beneath the simulation room, following which it is brought back by the fan into the simulation room creating one closed air flow pipe with the direction of movement depicted by grey arrows.

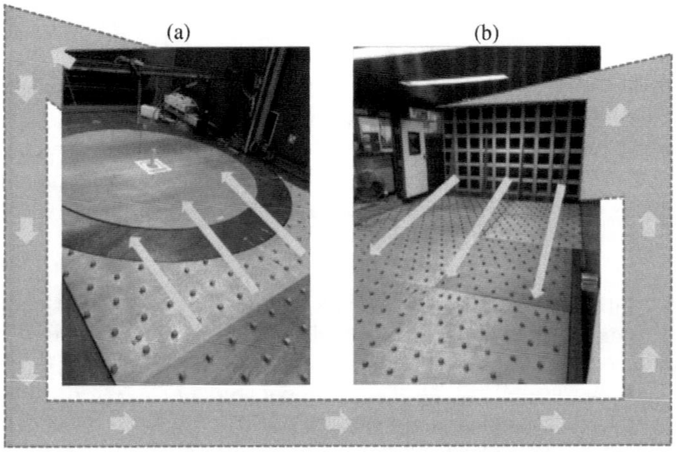

Figure 3. The boundary layer wind tunnel in HKU: (a) the rotatable round plate for fixing target models and (b) floor roughness elements in wind tunnel.

Two pressure sensors connected to a pitot static tube respectively, and they were employed to measure wind factors, such as wind speed, direction and pressure. One sensor and tube unit was installed on the moveable frame above the model, whilst the other one was fixed onto a further frame ahead of the models. This category of sensor and tube unit projects an "L" shape and it can only move vertically and horizontally when controlled by the external control panel. The wind speed can be monitored from the control panel that is also able to set up inlet wind velocities.

The shielding effects of various configurations of surrounding medium-rise buildings in the same wind tunnel laboratory have been tested in HKU [4]. In their research, the vertical profiles of mean wind speeds and turbulence intensities have been measured at the centre of the turntable for the simulated wind, for two terrain types: open land and suburban. Both the terrains have employed the mean wind profile following the power law with power exponent =0.13 and 0.20, respectively. Due to the geographical location (the north of Sheung Wan is sea) and urban morphology (located in high-dense tall building district), the suburban land terrain is chosen to describe Sai Ying Pun and the power exponent 0.20 will also be applied in the latter CFD simulation.

2.2. Procedure

The target geometric scale is 1:300; therefore, the tallest building with an approximate height of 150m will be 50cm in model-form. It is important to note that East and North are the two prevailing wind directions with the mean inlet wind velocity as 6m/s for both wind tunnel tests and CFD simulations.

IOP Conf. Series: Materials Science and Engineering **264** (2017) 012014 doi:10.1088/1757-899X/264/1/012014

3. CFD simulation

3.1. Set-up of the Simulation

The target site area is approximately 260 * 450 * 160m (W * L * H), while the whole computational domain size for this validation experiment is 2000 * 2000 * 400m. In this paper, considering software limitations, an adaptive meshing method has been employed in view of its ability to obtain increasingly accurate data on wind environment at the pedestrian level, which can, subsequently, utilise the computational resources efficiently too. In addition, three more layers are arranged below the evaluation height (1.5m above ground level, roughly human head/chest height where the thermal environment is sensed) and the maximum grid size ratio is set to 1.2. There are three levels of cells: In total, the domain is divided into 7,776,000 cells. Furthermore, the grid distribution is established based on the recommendation by the Architectural Institute of Japan (AIJ) [5], whereby 10 cells correspond to a cube root of building volume, which means the footprint of each building is divided into at least 10 cells to increase the accuracy of results. In this condition, the smallest size of cell in the interested with the most density area is about 1 * 1.2 * 0.5m (W * L * H).

3.2. Parameter settings

The boundary conditions, turbulence model and other boundary parameters have been considered according to relevant research [6-7]. For the purposes of this research, the CFD simulations have applied the 3D steady RANS equations and the standard k-ε turbulence model provides closure. A basic algorithm is used to stabilise pressure velocity-coupling. For the convection and viscous terms of the governing equations, pressure interpolation is the second order utilised.

3.3. Results

The wind flow can be influenced by objects as well as building patterns and orientation. Generation of high-pressure zones by wind deflection on windward facades is a well recognised phenomenon, while low-pressure zones are attributed to wind separation over the leading sharp edge and leeward façade. Indeed, wind environments with high velocity and turbulence may exist in the leeward and particular attention should be paid towards tackling such occurrences.

Simulation results at the pedestrian level performed using FlowDesigner with two prevailing wind inlet directions are presented in figure 4. Likewise for both wind directions, the central space is mostly wind-shaded (areas with blue dashes) and may experience low air movement. Surrounding buildings enclose these spaces and the corridors linking to main streets may be too narrow. Consequently, the majority of separate airflow over the wind-ward buildings did not enter these central spaces. On the other hand, the highest wind velocities are present in the outdoor spaces (purple dashes) for both wind directions as these areas face the inlet winds directly allowing the airflow to penetrate these narrow areas liberally. Comparison of the wind environment depicted by the green dashes indicates that the average wind speed and turbulence levels are higher in figure 4b than figure 4a due to there being increased airflow at the pedestrian level when the wind is from east, allowing wind to pass through the aisles between buildings. Furthermore, by analysing the distinction in wind pressure for both variables, the longer windward facades by buildings and podiums encourage greater wind deflection.

IOP Conf. Series: Materials Science and Engineering **264** (2017) 012014 doi:10.1088/1757-899X/264/1/012014

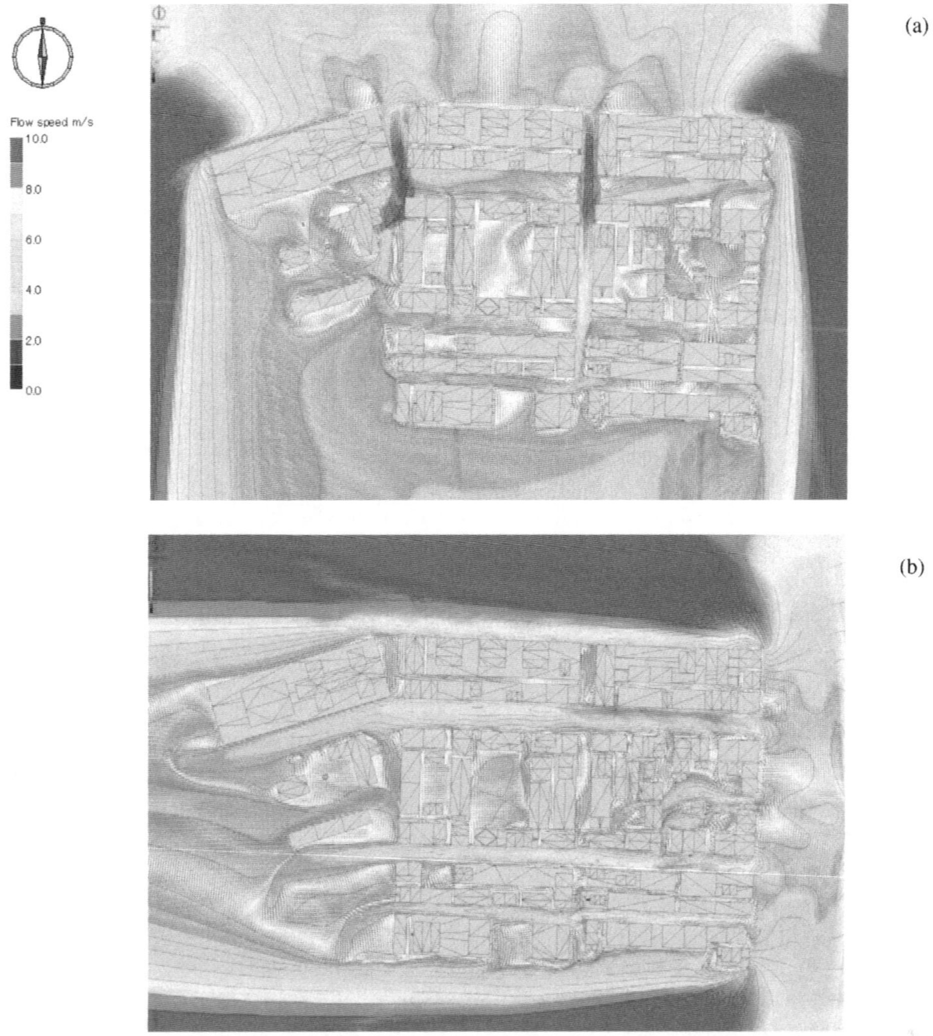

Figure 4. CFD simulation results in Sai Ying Pun: (a) north wind inlet and (b) east wind inlet.

4. Results comparison and discussion between wind tunnel tests and CFD simulations

For both wind tunnel tests and CFD simulation experiments, the mean wind velocity measurements were analysed at an equivalent full-scale height of 1.5m and a scaled height of 0.5cm in wind tunnel tests. The positions where comparisons of pedestrian level wind speeds were made are detailed in figure 5. The majority of these locations were selected after careful consideration during the experiment, whereby two positions (locations 4 and 9) encompassed the central aisles and three positions (locations 7, 10 and 11) covered the courtyard region.

There were two monitoring sensors were used during each wind tunnel test; one was always set ahead of the first building facing the incoming wind with a distance of 20cm and a height of 30cm from ground level in scaled, which equals to 60m and 90m respectively. This sensor was applied to ensure the inlet wind speed was keeping at the level of 6m/s. Meanwhile, another remotely operated sensor measured the wind speed of selected points. Once the wind speed at all thirteen positions was recorded with the north wind the building models were rotated 90° anticlockwise for the subsequent simulation.

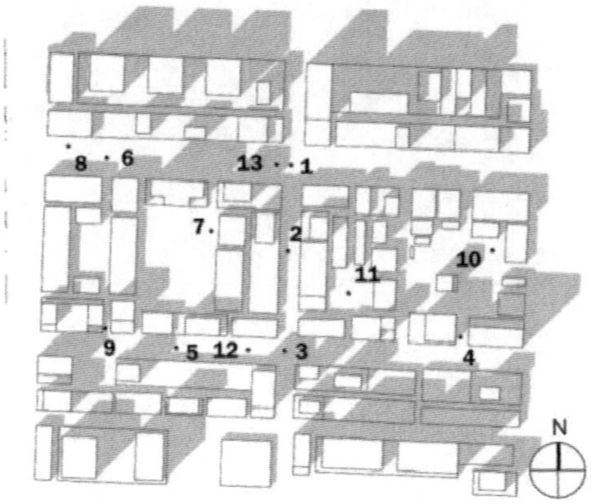

Figure 5. Positions in the site for comparing the results of wind tunnel tests and numerical simulation.

As shown in table 1, discrepancies in the findings with both inlet wind directions show an equivalent order of magnitude. Location 12 shows the maximum inconsistency between each wind direction variable, which is greater than approximately 200%. Conversely, the deviations are within 20% for most of the other locations. The maximum difference tends to occur within the courtyard areas (Locations 7 and 11), while the lowest variation arises nearest to the building corners (Locations 4, 9 and 10). This divergence may be explained by limitations of CFD in modelling complicated air flow patterns, including those in re-circulating regions. Alternatively, it may be caused by errors made when taking measurements on account of interference by sensors during wind tunnel experiments. In addition, the roughness of model surface in the wind tunnel is not as smooth as CFD, and it may reduce the level of details when meshing the model in CFD.

Table 1. Mean wind speeds with north and east wind in wind tunnel and CFD.

Location	North wind speed (m/s)			East wind speed (m/s)		
	wind tunnel reference wind velocity	wind tunnel	CFD	wind tunnel reference wind velocity	wind tunnel	CFD
1	6.1	7.06	8.16	6.39	5.72	6.54
2	6	6.01	6.3	6.44	2.95	1.78
3	6.08	2.28	2.5	6.38	4.83	5.76
4	5.96	3.78	3.61	7.24	3.35	3.12
5	5.93	3.62	3.36	6.61	5.52	6.15
6	6.12	4.2	3.65	6.45	6.45	6.28
7	6.23	3.97	3.14	6.66	1.75	0.54
8	6.14	2.75	3.83	6.42	5.97	5.43
9	6.01	3.55	3.39	6.4	3.12	2.76
10	6.07	2.89	2.7	6.87	0.76	0.9
11	6.02	3.32	6.43	6.69	1.05	2.31
12	5.87	2.53	2.12	6.29	4.63	4.97
13	6.09	2.05	2.24	6.46	5.81	6.2

Therefore, from the ratio between the measured wind speed and the reference wind speed at each selected location in wind tunnel tests and CFD simulation shown in figure 6, the great majority are between 0.8 and 1.3, which is suggesting that application of the standard k − ϵ model in FlowDesigner can predict accurately the pedestrian-level wind environment.

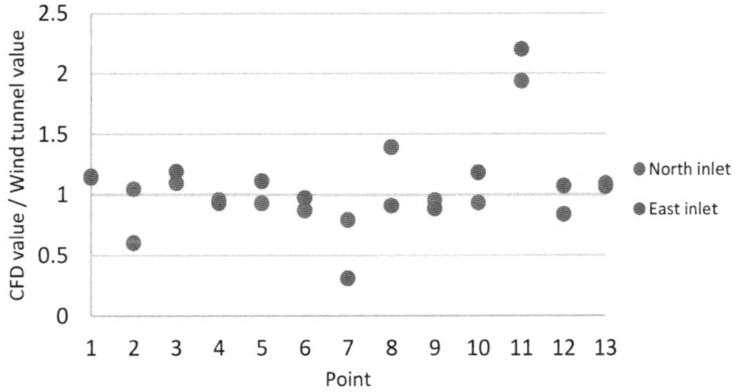

Figure 6. Wind speed ratio of CFD simulation and wind tunnel tests.

5. Summary

This paper has addressed the experimental and computational assessment of pedestrian outdoor wind environments neighbouring a series of tall buildings in Sai Ying Pun, Hong Kong. Local inhabitants' living conditions and natural wind statuses in the urban environment have been introduced. Therefore, the buildings have been used in wind tunnel tests and CFD to analyse the outdoor ventilation. The existing problems are the contradiction between planning control and commercial interests, the incongruity between public interest and individual interests, the ambiguity between society, environment and economic benefit is the long-term discussion topic in urban renewal. Integrating the results to the realistic state can demonstrate that, the elderly may have to use some narrow streets with high velocity, followed by dangerous, and citizens may also have to apply the courtyards with stagnant wind that is not beneficial to the healthy air change. The important way to deal with these contradictions is co-ordination, which means with regard to urban renewal and transformation movement in such high-density and high rise district, under the balance of society and economy, improving environmental comfort, health and safety in advance is the key strategy to make Hong Kong become environmental-friendly and sustainable. In future research about local urban renewal, elevating the podium of these tall buildings for allowing more wind to pass through the calm zone could be studied to improve the air exchange rate, especially at the pedestrian level.

6. References

[1] Yung, E. H. K., Zhang Q., Chan, E. H. W., 2017 Underlying social factors for evaluating heritage conservation in urban renewal districts. *Habitat International* 66 135-148

[2] Ng, E. 2009 Policies and technical guidelines for urban planning of high-density cities – air ventilation assessment (AVA) of Hong Kong. *Building and Environment* 44 1478-1488

[3] Ng, E. 2010 *Designing for urban ventilation* In: NG, E. (ed.) Designing high-density cities (the UK and USA: Earthscan)

[4] Lam, K. M., Wong, S. S. Y., To, A. P. & Fong, Y. F. 2013 Shielding Effects on A Tall Building from Various Configurations of Surrounding Medium-rise Building *The Eighth Asia-Pacific Conference on Wind Engineering* Chennai, India: Research Publishing

[5] Tominaga, Y., Mochida, A., Yoshie, R., Kataoka, H., Nozu, T., Yoshikawa, M. & Shirasawa, T. 2008 AIJ guidelines for practical applications of CFD to pedestrian wind environment around buildings *Journal of Wind Engineering and Industrial Aerodynamics* 96 1749-1761

[6] Mochida, A., Tabata, Y., Iwata, T. & Yoshino, H. 2008 Examining tree canopy models for CFD prediction of wind environment at pedestrian level *Journal of Wind Engineering and Industrial Aerodynamics* 96 1667-1677

[7] Hong, B. & Lin, B. 2015 Numerical studies of the outdoor wind environment and thermal comfort at pedestrian level in housing blocks with different building layout patterns and trees arrangement *Renewable Energy* 73 18-27

Acknowledgments

This research is funded by the National Natural Science Foundation of China (Grant No.51578378), the Special Funds for State Key R&D Program during the 13th Five-year Plan Period of China (Grant No.2016YFC0702104), the Sino-German Scientific Research Program (Grant No.GZ1162) and Science and Technology Commission of Shanghai Municipality (Grant No.16dz1206502).

IOP Conf. Series: Materials Science and Engineering **264** (2017) 012015 doi:10.1088/1757-899X/264/1/012015

Strength of laser welded joints of polypropylene composites

V Votrubec[1], P Hisem[1], L Vinšová[1] and V Lukášová[1]

1 Researcher, Department of Computations and Modelling, VÚTS, a.s., Liberec, Czech Republic
E-mail: vlastimil.votrubec@vuts.cz

Abstract. This paper deals with experimental tests of laser welded polypropylene composites. Polymers, such as polypropylene, are often filled with fibres in order to increase their mechanical properties. The welding procedure can also influence material properties nearby weld joints. Therefore the strength of weld joints is lower than strength of primary materials. This effect is proved by realized shear tests. Polymer specimens were filled with 20 % and 40 % of glass fibres and all possible combinations of specimens were welded for experiments. There is also discussed influence of volume fraction of glass fibres in polypropylene on the strength of weld joint.

1. Introduction
Nowadays, laser welding is common method of jointing polymers and its importance is increasing. Generally, welding of plastics is the process of jointing parts through heating in order to melt and fuse the polymer at the interface [1]. Advantages of laser welding to conventional welding methods are localized heat input to the joint interface, reduced welding flash, weld seams of high mechanical strength and quality [2]. Also the settings of welding conditions like speed of welding, laser power and focal distance are easy. So the repeatability of potential measurement is high.

1.1. Motivation
Mechanical properties of polymers compared to metals are worse. The value of material strength, yield point and stiffness is much lower but the enhancement of these properties can be achieved by adding fibres, fabrics or particles. Welding of reinforced composites is more complicated than welding of pure polymers but it is still possible. The effort of researchers is mainly aimed at welding process itself, welding conditions and possibilities of various material jointing [3, 4] whereas area of weld joint strength and material properties is not so inspected. Papers concerning on reinforced polypropylene mostly have to do with carbon fibres [5].
The main purpose of this article is to describe the influence of weld joint on total strength of various jointed materials. Material chosen for shear testing on testing machine was glass fibre reinforced polypropylene with 0, 20 and 40 % amount of fibres.

1.2. Materials properties
Jointing of plastics by laser welding and shear tests as well have certain constraints and specifics. One of the pair of the welded polymer needs to be transparent and one absorbing, usually clear (or natural) and black. Although jointing of two transparent polymers is possible [6, 7], common method (used in this research) is doping the polymer with absorbing additive. These additives, e.g. carbon black, partially change material properties (see Table 1).

The measurement of mechanical properties of both natural and black polymers had to be conducted before the shear tests. Afterwards it was possible to compare the tensile strength of origin basic material and strength of weld joint. The specimens were cut out of an injected plate. Material experiments were performed on testing machine as standard tension tests and only the strength of material was important for following evaluations. The material strength of used glass fibres reinforced polypropylene is in Table 1.

Table 1. Materials used for jointing by laser welding and their tensile strength. The letter "N" means natural (clear) material, letter "B" means black absorbing material and number denotes the volume fraction in % of glass fibres.

Material	Tensile strength [MPa]	Material	Tensile strength [MPa]
PP N0	33	PP B0	35.7
PP N20	60.4	PP B20	58.2
PP N40	60.6	PP B40	69

2. Measurement and welding

The shear test was proposed as a tensile test of two welded overlapped pieces of material. None of European Standard describes this type of experiments so proper methodology was specified. The scheme of welded specimens is in Fig. 1. Dimensions of specimens were 130×20×2 mm and the lap length was 20 mm. The weld joint was continual across the specimen and its width was 1 mm.

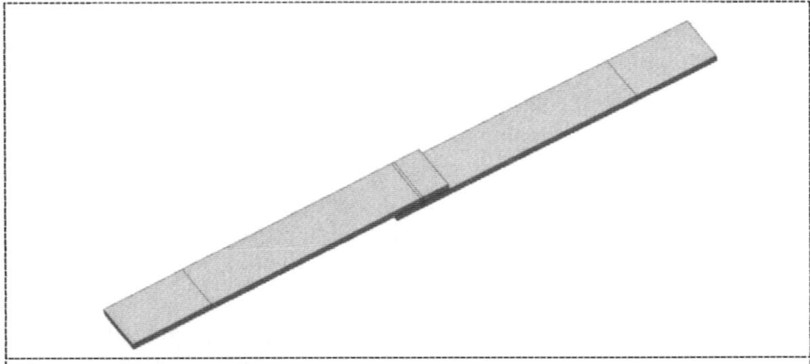

Figure 1. The scheme of welded materials prepared for shear test.

The end areas were clapped into grips of testing machine Instron and loaded by increasing force till the rupture. The velocity of crosspiece was 1 mm/min. The main result was the maximal force on the tensile curve, which is the strength of the weld joint. Then the area of joint weld of each specimen was measured by the image processing and all data were statistically processed and evaluated. Nine combinations of materials were welded for the teste, see Table 2.

Table 2. Combinations of welded materials for shear tests. The letter "N" means natural material, letter "B" means black material and number denotes the volume fraction in % of glass fibres.

Material	Material	Material
PP N0-B0	PP N20-B0	PP N40-B0
PP N0-B20	PP N20-B20	PP N40-B20
PP N0-B40	PP N20-B40	PP N40-B40

Specimens for shear tests were welded on laser station LM05/05P.W. Settings of welding parameters were quite demanding because the weld joints needed to be thin. The presence of glass fibres also complicated the welding process as the laser beam was dispersed by the fibres. Dimensions of the weld joints weren´t optimal (they were small) but larger area of the joints would cause fracture in the origin material during the test, not in the weld joint. On the Fig. 2 there is both welded parts pressed to the base by a tool. This tool contains of four pneumatic cylinders ensuring adequate contact pressure on specimens. In the middle of the tool there is longitudinal hole for laser beam welding the specimen. Welding parameters were unique for each combination of welded materials to achieve good weld joint. The higher volume fraction of fibres in polypropylene, the higher welding velocity had to be used.

Figure 2. Work desk of laser station LM05/05P.W during the preparation on welding.

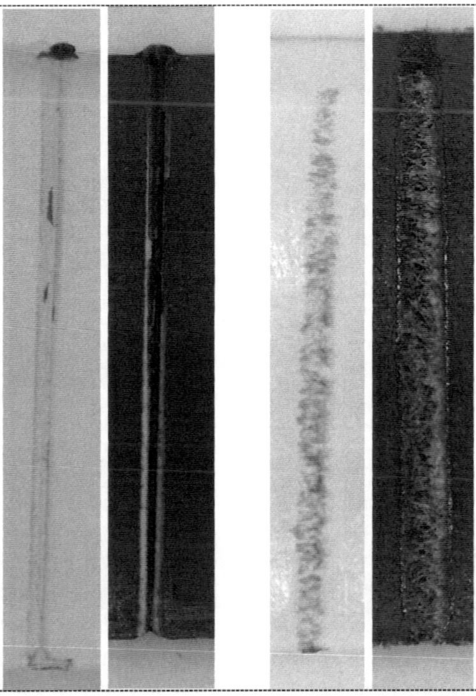

Figure 3. Separated parts of welded specimens PP N0-B0 (left) and N20-B40 (right) after the test under the microscope.

3. Measurement evaluation

Some specimens cracked too soon and the measured maximal force was very low. It was probably caused by lacking welding, awkward manipulation or other mistake. These specimens were taken out of the statistical processing, there were less than 10 specimens total. The measurement evaluation is in following Table 3 and photographs of broken specimens on the Figure 3.

Table 3. Statistical processing of experimental data obtained by shear test of welded polypropylene composite specimens.

	Maximal force [N]	Weld area [mm^2]	Maximal stress [MPa]
PP N0-B0	246 ± 16	14.6 ± 1.5	16.9 ± 1.1
PP N0-B20	254 ± 11	14.5 ± 1.0	17.6 ± 1.2
PP N0-B40	290 ± 8	19.8 ± 0.8	14.7 ± 0.5
PP N20-B0	319 ± 22	18.6 ± 1.4	17.2 ± 1.0
PP N20-B20	314 ± 39	17.9 ± 1.6	17.6 ± 1.4
PP N20-B40	504 ± 56	24.6 ± 1.4	20.4 ± 1.9
PP N40-B0	277 ± 50	18.6 ± 2.6	14.9 ± 1.5
PP N40-B20	476 ± 57	24.2 ± 2.5	19.6 ± 0.7
PP N40-B40	702 ± 28	36 ± 2.0	19.5 ± 1.2

For each combination of welded materials the mean value of measured maximal force and evaluated weld area and maximal stress are listed. The stress is a fraction of force and weld area. The mean value is appended with the value of 95% confidence interval.
Already quick view to the Table 3 indicates that presence of glass fibres in polypropylene doesn´t increase the strength of weld joints, decisively not proportionally to the strength of welded materials. According to the results (values of maximal stress), the dependence of weld joint strength on volume fraction of fibres can´t be done. The stress values form the Table 3 are shown in Fig. 4 for better understanding.

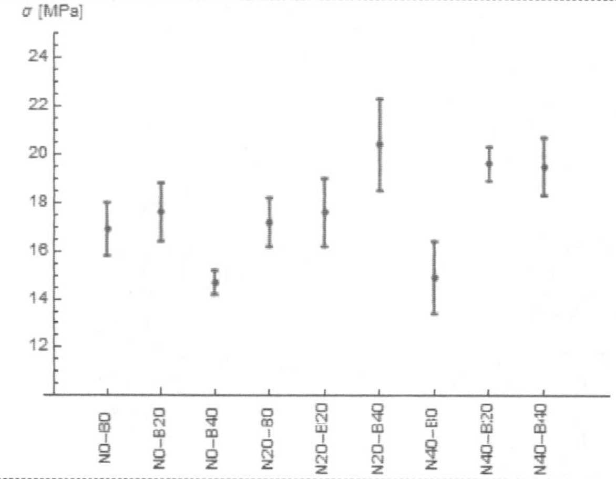

Figure 4. Mean value and 95% confidence interval of evaluated maximal stress of welded PP composites.

The measured force and evaluated stress corresponds to the shear stress. Shear strength is usually approximately half compared to a material tensile strength. So the weld joints strength of non-filled polypropylene match the expectations, see the Fig. 5. The strength of reinforced PP was much higher but the weld joint strength still remains very similar. It seems the weld joint strength depends mainly on material strength of the basic non-filled material. The reason why the fibres don´t increase the weld joint strength similarly to material strength is meanwhile unknown.

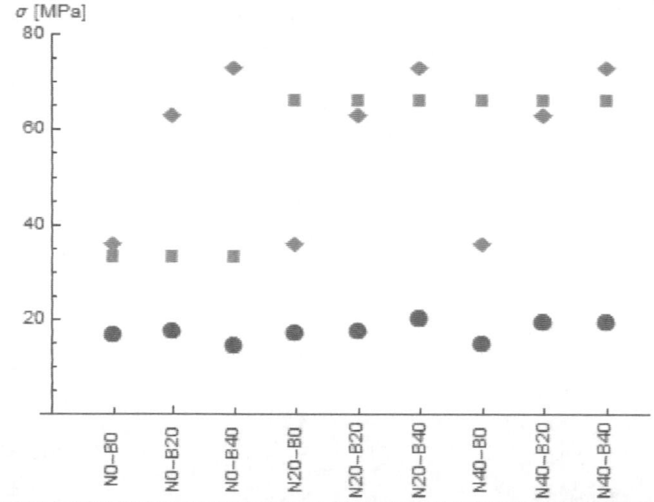

Figure 5. Comparison of weld joint strength (disk) and material. strength (square – natural PP, rhomb – black PP).

4. Conclusion

The weld joint strength was not surprisingly lower than strength of welded materials. This was proved by shear tests on testing machine. Quite unexpected result was little influence of glass fibres on weld joint strength. The weld joint strength remains about the same value for all welded material combinations. The weld joint strength of non-filled materials was approximately half in comparison with strength of origin material.

5. References

[1] Šercer M and Raos P 2010 Joining of Plastics and Composites *Welding Engineering and Technology* ed. S Kralj (EOLSS Publishers)
[2] Klein R. 2012 *Laser Welding of Plastics* (Germany: Wiley-VCH GmbH & Co. KGaA)
[3] Jaeschke P, Wippo V, Suttmann O and Overmayer L 2015 *Journal of Laser Applications* 27(S2)
[4] Ghorbel E, Casalino G, Abed S 2009 *Material and Design* 30(7), p 2745
[5] Bates P J, Tan S, Zak G and McLeod M 2009 *Composites Parts A: Applied Science and Manufacturing* 40(1), pp 28-35
[6] Boglea A, Rosner A and Olowinsky A 2010 *Proc. Int. Congress on Applications of Lasers & Electro-Optics (Orlando)* vol 103, p 567
[7] Ruotsalainen S, Laakso P and Kujanpaa V 2015 *Physics Procedia* 78, p 272

Acknowledgments

This paper was written as a part of a project LO1213, supported by the Czech Ministry of Education.

IOP Conf. Series: Materials Science and Engineering **264** (2017) 012016 doi:10.1088/1757-899X/264/1/012016

Optimization of heat treatment parameters for additive manufacturing and gravity casting AlSi10Mg alloy

L Girelli[1], M Tocci[1], L Montesano[1], M Gelfi[1] and A Pola[1]

1 Department of Mechanical and Industrial Engineering, University of Brescia, via Branze 38, 25123, Brescia (Italy)
E-mail: l.girelli005@unibs.it

Abstract. Additive manufacturing of metals is a production process developed in the last few years to realize net shape components with complex geometry and high performance. AlSi10Mg is one of the most widely used aluminium alloys, both in this field and in conventional foundry processes, for its significant mechanical properties combined with good corrosion resistance. In this paper the effect of heat treatment on AlSi10Mg alloy was investigated. Solution and ageing treatments were carried out with different temperatures and times on samples obtained by direct metal laser sintering and gravity casting in order to compare their performance. Microstructural analyses and hardness tests were performed to investigate the effectiveness of the heat treatment. The results were correlated to the sample microstructure and porosity, analysed by means of optical microscopy and density measurements. It was found that, in the additive manufactured samples, the heat treatment can reduce significantly the performance of the alloy also because of the increase of porosity due to entrapped gas during the deposition technique and that the higher the solution temperature the higher the increase of such defects. A so remarkable effect was not found in the conventional cast alloy.

1. Introduction

Additive manufacturing (AM) is a production process whereby products are made through a layer by layer melting of a metallic powder (or wire) with a focused energy source [1-2]. This manufacturing method is largely used for aircraft and automotive fabrication [3].

Additive manufacturing technology has revolutionized components manufacturing and logistics, allowing the production on demand, the reduction of energy consumption [4] and the manufacture sustainability [5]. It provides several advantages such as a reduction of material used [6-7], unrivalled freedom of design [1, 5-6, 8-10], elevated quality [11] and a lightening of the final component [12]. It is also important to consider the lead time reduction [1-2, 13], the possibility to realize shapes specifically designed on the requests of the customer [11] that cannot be obtained by traditional production methodologies [6, 14]. Nowadays, different techniques are available for AM of metals, like: laser beam melting, laser metal deposition or electron beam melting. In particular, the laser beam melting process is also known as selective laser melting (SLM), direct metal laser sintering (DMLS), laser metal fusion etc., mainly depending on the trademark of the machine producer [1]. The common metals for AM are: some Ti, Ni, Al and Co alloys and some steels [1].

Aluminium alloys have been considered as an alternative material to replace steels in automotive industry, owing to increasingly strict standards with respect to fuel efficiency, air pollution, recycling and safety [15-16]. For this reasons, today, many automotive components are made in aluminium alloy, such as engine blocks, cylinder heads, wheels, suspension links, engine heads, pumps, callipers, etc.

[17-18]. Many of these parts are typically produced by foundry processing, like gravity or low pressure die casting, with Al-Si alloys containing age hardenable elements [17]. In these alloys, in fact, the small addition of Mg (in the order of 0.3-0.5 wt.% Mg) induces hardening of the aluminium alloy by forming Mg_2Si precipitates upon solution treatment followed by natural or artificial ageing (T4 or T6 respectively) [17].

Lately, due to the above-mentioned advantages, also additive manufacturing has been applied for the production of high performance aluminium parts [1]. In particular, AlSi10Mg alloy is one of most widely used aluminium alloys in AM, as well as in foundry processing, because of its high mechanical properties, low thermal expansion and good corrosion resistance [17]. Typically, this alloy undergoes T6 treatment in order to enhance its performance. Some authors have already reported the temperatures for both solution and ageing treatment, ranging between 500-545°C and 150-180°C respectively [19-20]. The durations are normally based on studies carried out on conventionally cast parts, which are characterized by different microstructure compared to AM, thus showing a different response. Aboulkhair et al., for instance, reported that a conventional T6 performed on selective laser melting (SLM) AlSi10Mg results in softening the materials, raising the question of the validity of conventional heat treatments on AM parts [20]. Similarly, Thijs et al. showed that the as-built SLM AlSi10Mg alloy presents a hardness comparable to the high-pressure die-cast alloy in aged condition [21].

Notwithstanding these findings, SLM AlSi10Mg parts are usually heat treated. This happens because, after solution, quenching and ageing treatment, the former anisotropy in the microstructure, caused by the layer by layer manufacturing, is dissolved [1] and an improved ductility is gained [22]. Hence, as already outlined by some authors [19-20], it is fundamental the definition of a new set of heat treatment procedures, specifically tailored for AM parts.

The aim of this work is to analyse systematically the effect of different temperatures and durations, of both solution treatment and artificial ageing, on hardness and microstructure evolution of AM AlSi10Mg parts, considering in particular the combined effect of porosity. The same investigation was performed on AlSi10Mg gravity cast parts to compare the different behaviour.

2. Experimental procedure

A commercially available AlSi10Mg powder was used in this study for the production of samples by direct metal laser sintering technique (DMLS). The manufacturer used proprietary process parameters to produce the parts.

Cast samples were obtained from a component produced by gravity casting (GC) in a permanent mould. Attention was paid to cut the samples from areas with the same section width in order to ensure the same solidification condition and the same microstructural characteristics. The samples chemical composition is shown in Table 1 for both these manufacturing techniques.

Table 1. Chemical composition (wt. %) of studied alloy.

	Si	Mg	Fe	Cu	Mn	Ni	Zn	Ti
AlSi10Mg	9 – 11	< 0.55	< 0.55	< 0.05	< 0.45	< 0.05	< 0.12	< 0.15

For both the considered production processes, some samples were tested in as cast condition, while the others were heat treated in a laboratory furnace. Several heat treatment conditions were applied by changing temperature and time. Particularly, solution treatment was performed at 510°C and 540°C for 6 hours. Then, samples were water quenched at 65°C. After each solution treatment and quenching, half samples were aged at 160°C, while the others at 170°C. Ageing time ranged from 1 to 10 hours in order to obtain a complete ageing curve. Between solution and ageing treatment, samples were kept at -20°C in order to avoid natural ageing.

The specimens were grinded and polished with waterproof paper and diamond cloths up to mirror finishing. Microstructural observations were carried out by optical microscope Leica DMI 5000 M equipped with LAS software; DMLS microstructure was also analysed after Keller etching for 20 s.

The density of the all samples was measured by hydrostatic weighing method using a proper balance (Gibertini E42-B) and holding system both before and after heat treatment. According to this method, samples were weighted in two different media, air and water, and the sample density ρ_{sample} was calculated with the following formula:

$$\rho_{sample} = \rho_{water} \cdot \frac{W_{air}}{W_{air} - W_{water}} \quad (1)$$

where W_{air} and W_{water} are the weight of the sample in air and water, respectively, and ρ_{water} is the water density.

Finally, material hardness was measured by Brinell method with a Galileo Ergotest Comp25 apparatus (LTF Galileo Italy) applying a load of 613 N for 15 s and using a ball of 2.5 mm of diameter as indenter. The measure was repeated 5 times for each sample. Hardness values were used to evaluate the effectiveness of the heat treatments. In addition, the influence of porosity on this mechanical property was investigated in order to completely characterize the materials according to the specific manufacturing process.

3. Results and discussion

In Figures 1-2 the microstructure of DMLS and GC samples is reported in as-produced condition, as quenched at 510°C and after ageing at 160°C for 2 hours, as an example of the analysed conditions.

GC samples show the typical microstructure of casting products, characterized by the dendritic Al-rich phase (light grey in the images) surrounded by the modified eutectic, where some micro-shrinkage and hydrogen porosities were detected (Fig. 1a). The solution treatment, as predictable, induces spheroidization and coarsening of the Si particles as visible in Fig. 1b. The ageing does not remarkably affect the microstructure at this magnification. Similar evidences were gathered in the case of solution performed at 540°C and of ageing at 170°C.

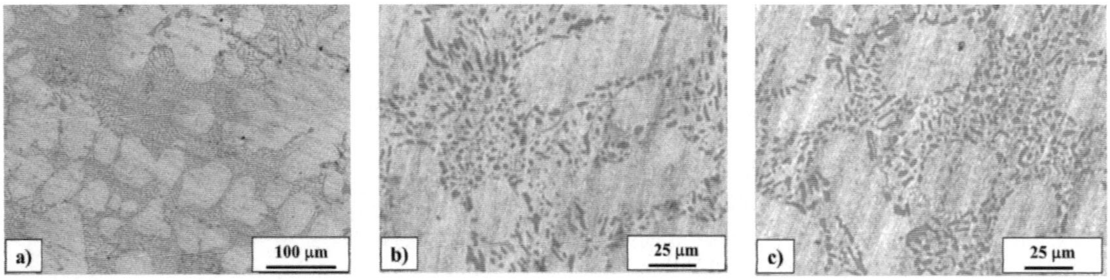

Figure 1. Microstructure of GC samples in as-cast (a), quenched (b) and 2h-aged condition (c).

Concerning the DMLS samples, the microstructure reveals the typical pools of semi-circular section along the building direction (Fig. 2a) and elongated in the deposition plane, according to the laser beam scanning paths. At higher magnifications, a very fine Al-rich cellular grain can be observed as well as very fine Si particles at the borders, as extensively described in literature [1, 23-24]. During the solution treatment, Si forms larger particles in the matrix (Fig. 2b), preferentially at the former pools boundaries, that now are no more visible [1]. Small porosities can be frequently detected in the DMLS samples, which can be due to the powder quality, hydrogen absorption, oxide inclusions, melt splashing, Marangoni flow during laser scans, etc. [4, 25-29]. The microstructure of DMLS samples after aging does not revealingly change (Fig. 2c).

A slight coarsening of the microstructure can be observed increasing the solution temperature as well as the ageing time, in agreement with other findings [22].

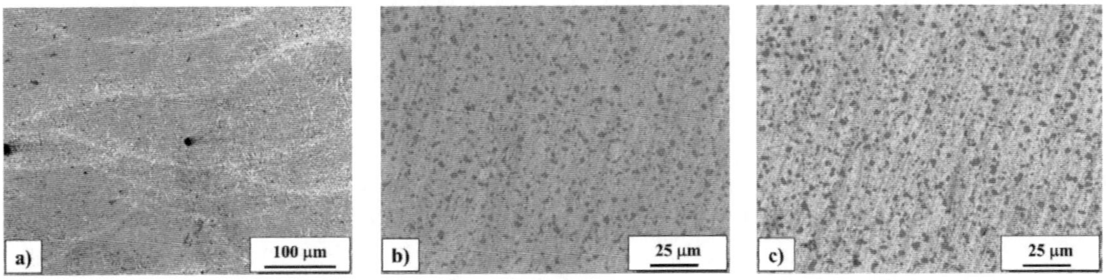

Figure 2. Microstructure of DMLS samples in as-built (a), quenched (b) and 2h-aged condition (c).

As-produced parts are characterized by very different performance compared to the heat treated ones (Fig. 3). In particular, the as-built alloy shows very high hardness, with a mean value of 125.20±1.05 HB, while the as-cast one results 63.37±1.55 HB. This superior resistance of the as-built material is related to the grain refinement and nano-sized eutectic Si obtained as a consequence of the fast cooling [22]. Additionally, the alloy experiences a "self quenching" phenomenon because of the layer by layer deposition process. In fact, the rapid cooling and the subsequent re-heating during the layer deposition can act like a T6 treatment, promoting the precipitation of Mg_2Si [30].

The subsequent solution treatment equalizes the hardness of the DMLS and GC alloys, independently from the tested temperature. In detail, after quenching, the hardness of the GC alloy remains almost the same as before; on the contrary, that of the DMLS alloy drops to about a half of the initial value.

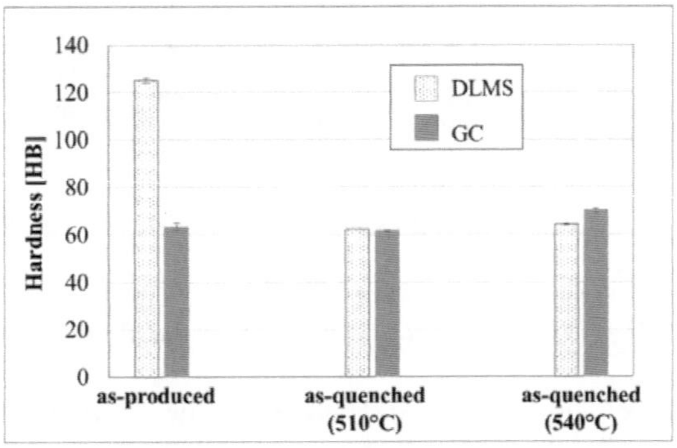

Figure 3. Brinell hardness of as-produced and as-quenched samples.

This remarkable reduction in the DMLS alloy hardness is due to the dissolution of the previously formed precipitates combined with the possible increase of porosity defects.

To evaluate the effect of the porosity level on mechanical properties, the density of the samples was measured at each treatment condition. The as-built alloy shows a density of about 2.679 g/cm^3, comparable to that of the as-cast part (2.669 g/cm^3). After quenching, the density of the GC alloy remains almost the same for both the solution temperatures, as shown in Table 2. For the DMLS, when the solution treatment is performed at higher temperature, a reduction in density of about the 5% is obtained.

Table 2. Density of GC and DMLS alloys before and after quenching.

	GC [g/cm^3]	**DMLS [g/cm^3]**
As-produced	2.669 ± 0.006	2.679 ± 0.003
As-quenched (from 510°C)	2.672 ± 0.007	2.638 ± 0.004
As-quenched (from 540°C)	2.671 ± 0.008	2.558 ± 0.006

IOP Conf. Series: Materials Science and Engineering **264** (2017) 012016 doi:10.1088/1757-899X/264/1/012016

This remarks are further supported by the observation of the microstructure of samples in as-produced and quenched conditions. In fact, for GC samples, no significant increase of porosity can be identified, and both micro- and macro-porosities can be found (Fig. 4).

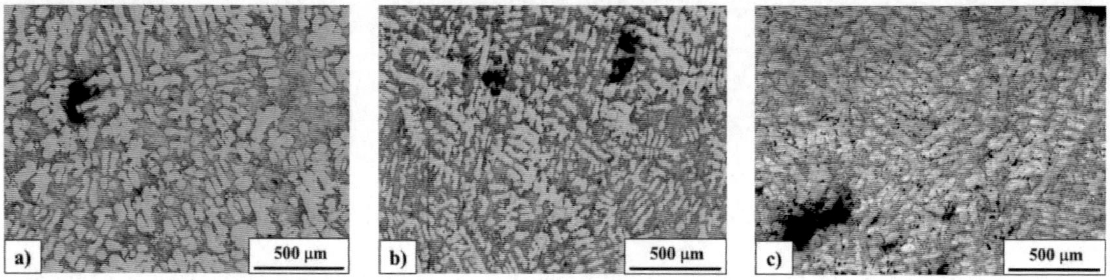

Figure 4. Micrographs showing different porosity of GC samples in as-cast (a), quenched after solution treatment at 510°C (b), quenched after solution treatment at 540°C (c) condition.

On the other hand, in Figure 5, it is possible to observe a remarkable increase in porosity after solution treatment of the DMLS samples, with a very high number of small pores, especially after treatment at 540°C, which is consistent with density measurements (Table 2).

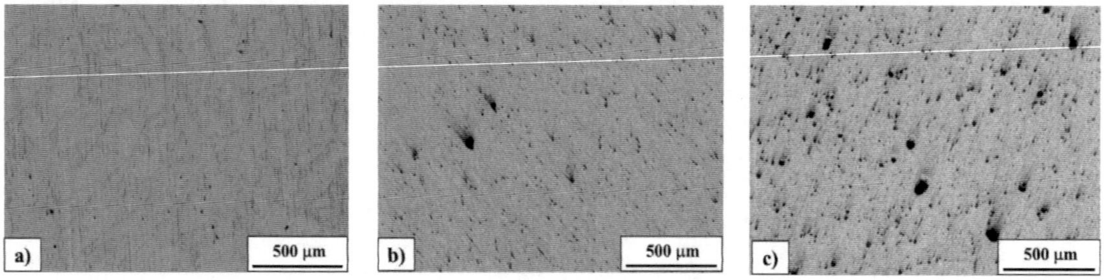

Figure 5. Micrographs showing different porosity of DMLS samples in as-built (a), quenched after solution treatment at 510°C (b), quenched after solution treatment at 540°C (c) condition.

An explanation of this decrease in density for DMLS alloy after solution treatment can be found if the characteristics of this innovative manufacturing route are taken into account. As already mentioned, the used technology allows the formation of small porosities in the as-built part. In fact, due to the laser beam, the liquid pool reaches very high temperatures, promoting the dissolution of gas, such as hydrogen coming from the moisture on the powder particle surface as well as dissolved in the powder itself [31]. This results in the formation of pores in the solidified alloy. Additionally, because of the fast solidification and cooling, it is enhanced the entrapment of gas atoms into the aluminium lattice, since they have no time to diffuse.

During solution treatment, the diffusivity of these atoms in Al is known to increase remarkably [32-33] in comparison with that at room temperature. This can lead to the growth of pores size and number, due to their enrichment of gas atoms previously present in solid solution in the bulk, and, therefore, to the measured decrease in density. In particular, between 510°C and 540°C, an increase of about 30% in diffusivity coefficient can be calculated [33], explaining the worse microstructure quality.

A similar pronounced mechanism of porosity increase with heat treatment is instead not possible for cast parts, where the solidification rate is significantly lower and allows the rejection of gas atoms from the metal during solidification itself, resulting in the well-known gas porosities.

Hardness ageing curves in Figure 6 show the different response to ageing treatment of the two alloys. Values for 0 hours of ageing correspond to the samples condition after solution treatment and quenching.

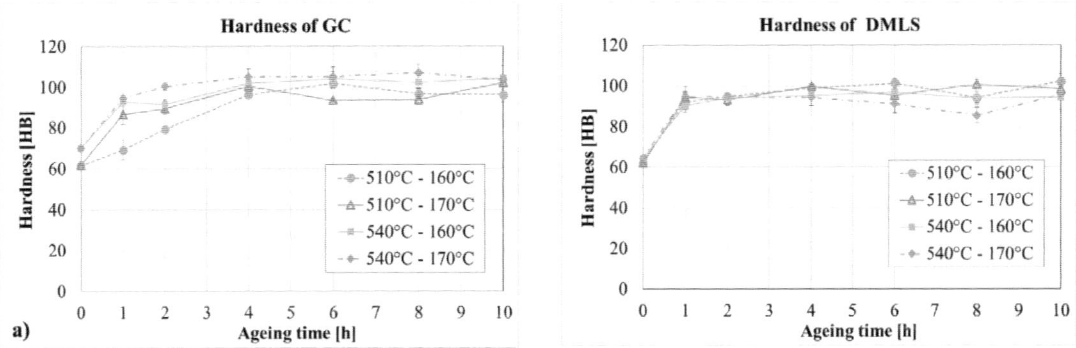

Figure 6. Brinell hardness ageing curves of GC (a) and DMLS (b) alloys.

In case of the GC alloy, the 510°C solution treatment results, in general, less effective than that performed at 540°C, requiring longer ageing time to reach comparable hardness values (Fig. 6a). For instance, after solution treatment at 510°C, a 6 hours ageing at 160°C is needed to obtain the same performance that can be reached after only 2 hours at 170°C when the solution is carried out at 540°C. The maximum hardness is achieved when a 540°C solution followed by a 170°C-8 hours ageing is done.

The ageing curves of the DMLS alloy appear slightly different (Fig. 6b). In particular, it can be seen that after 1 hour almost the same hardness is obtained, independently from the analysed treatment conditions. The highest hardness is reached after 510°C solution followed by 160°C-6 hours ageing, while the treatment at 540°C-170°C shows the lower values, independently from the ageing time.

Furthermore, it has to be noticed that the best heat treatment condition for the GC alloy is the worst for the DMLS one. In fact, the hardness at 540°C solution and 170°C-8 hours ageing treatment results 106.99±3.85 HB and 85.51±0.42HB for GC and DMLS respectively.

The reason of this different behaviour can be correlated to the presence of porosity defects, which were investigated by density measurements on aged samples for all the experimental conditions. Density values as a function of heat treatment parameters are shown in Figure 7.

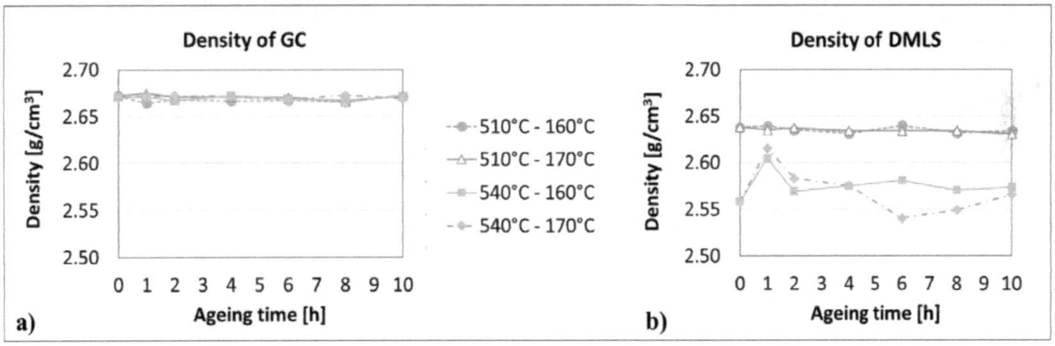

Figure 7. Density values of GC (a) and DMLS (b) alloys as a function of ageing time.

Considering GC samples (Fig. 7a), the density does not vary remarkably during ageing treatment in comparison with that of the quenched condition.

More interesting are the results for DMLS material (Fig. 7b). In fact, density values of aged samples after solution treatment at 510°C are quite constant for all the considered temperatures and durations. This implies that, in this condition, the ageing treatment does not induce a remarkable increase in porosity, with positive contribution to mechanical properties. This is in agreement with hardness data since the highest values were measured after solution treatment at 510°C (Fig. 6b).

Instead, as already discussed for quenched condition, noticeable lower density values were found for samples treated at 540°C for all the investigated combinations of ageing time and temperature.

Additionally, samples density is quite constant for treatment at 160°C, while in the case of 170°C, the values vary without a clear trend. It is believed that heat treatment at highest temperatures induces a non-uniform distribution of porosities, which results in this scattered behaviour.

4. Conclusion

In this study, it was investigated the effect of different temperatures and durations of T6 heat treatment on the properties of AlSi10Mg produced by means of gravity casting and direct metal laser sintering techniques.

In the as-produced condition, the hardness of DMLS part results the double of the GC one, as a consequence of the very fine microstructure and of a sort of self-treatment experienced by the alloy during the layer by layer production. After quenching, the hardness of both the samples appeared almost the same, with a strong reduction of performance of the DMLS alloy, compared to the as-built condition. This significant drop in hardness for the DMLS alloy is clearly due to the microstructural changes related to the treatment itself, but also to the remarkable increase in porosity. The evolution of porosity can be ascribed to the pores already present in the as-built component, combined with the atoms of hydrogen in solid solution in the aluminium lattice, unavoidably present because of the manufacturing method. The maintenance at high temperature during the heat treatment allows the diffusion of gas atoms and formation and growth of porosities; moreover, the higher the temperature, the higher the increase of such defects. In fact, the solution treatment performed at 540°C resulted deleterious for hardness and porosity. The density of DMLS decreases of 1.5% after solution at 510°C and quenching and of 4.5% after solution at 540°C and quenching, as a consequence of the increase in hydrogen diffusivity of about 30%. Microstructural analysis confirmed the density measurements. On the other hand, in the case of GC alloy, such reduction was not revealed.

Concerning the temperature and duration of the ageing treatment, it was shown by density measurements and microstructural analyses that they affect less the porosity, for both the studied alloys.

Furthermore, it was noticed that for the DMLS part none of the investigated temperature and duration of the treatment allowed to reach the same values of hardness measured on the as-built samples, supporting the theory of the influence of the porosity on mechanical properties.

Finally, it was observed that the optimized heat treatment parameters for the AlSi10Mg GC alloy result the worst for the DMLS one, showing the need of the definition of proper heat treatment procedures specifically tailored for AM parts.

5. References

[1] Herzog D, Seyda V, Wycisk E and Emmelmann C 2017 Acta Materialia 117 371
[2] Singh S, Ramakrishna S and Singh R 2017 Journal of Manufacturing Processes 25 185
[3] Murr L E 2016 Journal of Material Science & Technology 32 987
[4] Frazler W E 2014 JMEPEG 23 1917
[5] Ahn D G 2016 International Journal of Precision Engineering and Manufacturing - Green Technology 3-4 381
[6] Bhushan B, Caspers M 2017 Microsyst Technol 23 1117
[7] Petrovic V, Gonzalez J V H, Ferrando O J, Gordillo J D, Puchades J R B and Grinan L P 2017 International Journal of Production Research 49 1061
[8] Bikas H, Stavropoulos P and Chryssolouris G 2016 Int J Adv Manuf Technol 83 389
[9] Shirazi S F S, Gharehkhani S, Mehrali M, Yarmand H, Metselaar H S C, Kadri N Aand Osman N A A 2015 Science and Technology of Advanced Materials 16
[10] Guo N, Leu M C 2013 Front. Mech. Eng. 8 215
[11] Olakanmi E O, Cochrane R F and Dalgarno K W 2015 Progress in Materials Science 74 401
[12] Cooper D E, Stanford M, Kibble K A and Gibbons G J 2012 Materials and Design 41 226
[13] Cruz M F, Borille A V 2017 J Braz. Soc. Mech. Sci. Eng. 39 177

[14] Hirt L, Reiser A, Spolenak R and Zambelli T 2017 Advance materials 29
[15] Yan L Z, Zhang Y-A, Xiong B-Q, Li X-W, Li Z-H, Liu H-W, Huang S-H and Zhao G 2017 Rare Met. 36 550
[16] Kastensson Å 2014 Journal of Cleaner Production 66 337.
[17] Davis J R 1991 ASM Specialty Handbook: Aluminum and Aluminum Alloys ASM International
[18] Kaufman J G, Rooy E L 2004 Aluminum Alloy Castings: Properties, Processes, and Applications ASM International
[19] Aboulkhair N T, Tuck C, Ashcroft I, Maskery I and Everitt N M 2015 Metallurgical and Materials Transactions A 46 A 3337
[20] Aboulkhair N T, Maskery I, Tuck C, Ashcroft I and Everitt N M 2016 Materials Science & Engineering A 667 139
[21] Thijs L, Kempen K, Kruth J P and Humbeeck J V 2013 Acta Materialia 61 1809
[22] Li W, Li S, Liu J, Zhang A, Zhou Y, Wei Q, Yan C and Shi Y 2016 Materials Science & Engineering A 663 116
[23] Wu J, Wang X Q, Wang W, Attallah M M and Loretto M H 2016 Acta Materialia 117 311
[24] Cabrini M, Lorenzi L, Pastore T, Pellegrini S, Manfredi D, Fino D, Biamino S and Badini C 2016 Journal of Materials Processing Technology 231 326
[25] Louvis E, Fox P and Sutcliffe C J 2011 Journal of Materials Processing Technology 211-2 275
[26] Aboulkhair N T, Everitt N M, Ashcroft I and Tuck C 2014 Additive Manufacturing 1 77
[27] Maskery I, Aboulkhair N T, Corfield M R, Tuck C, Clare A T, Leach R K, Wildman R D, Ashcroft I A and Hague R J M 2016 Materials Characterization 111 193
[28] Ng G K L, Jarfors A E W, Bi G and Zheng H Y 2009 Appl Phys A 97 641
[29] Qiu C, Panwisawas C, Ward M, Basoalto H C, Brooks J W and Attallah M M 2015 Acta Materialia 96 72
[30] Jägle E A, Sheng Z, Wu L, Lu L, Risse J, Weisheit A and Raabe D 2016 JOM 68 943
[31] Weingarten C, Buchbinder D, Pirch N, Meiners W, Wissenbach K, Poprawe R 2015 Journal of Materials Processing Technology 221 112
[32] Ichimura M, Sasajima Y and Imabayashi M 1991 Materials Transactions 32 1109
[33] Hayashi S 1998 Jpn. J. Appl. Phys. 37 930

Acknowledgments
Authors wishing to acknowledge Mr. Domenico Cosio of Ghial S.p.A., for providing the casting samples, and Ms. Safaa Al Balushi of GUtech, for the support in samples preparation and analysis.

Tool wear of (Ti, Al) N-coated polycrystalline cubic boron nitride compact in cutting of hardened steel

Tadahiro Wada[1] and Hiroyuki Hanyu[2]

1 Nara National College of Technology, Department of Mechanical Engineering, Nara, Japan
2 OSG Corporation, Aichi, Japan
E-mail: wada@mech.nara-k.ac.jp

Abstract. Polycrystalline cubic boron nitride compact (cBN) is effective tool material for cutting hardened steel. In addition to coated high speed steel and coated cemented carbide that has long been used for cutting materials, more recently, coated cBN has also been used. In this study, to verify the effectiveness of the (Ti,Al)N-coated cBN, which is formed on the substrate of cBN by the physical vapor deposition method, the hardened steel was turned with the (Ti,Al)N-coated cBN tool at a cutting speed of 3.33, 5.00 m/s, a feed rate of 0.3 mm/rev and a depth of cut of 0.1 mm. Furthermore, the uncoated cBN, which was the substrate of the (Ti,Al)N-coated, was also used. The tool wear of the cBN tools was experimentally investigated. The following results were obtained: (1) The contact area between the rake face and the chip of the (Ti,.Al)N-coated cBN tool was smaller than that of the uncoated cBN tool. (2) The tool wear of the (Ti,Al)N-coated cBN was smaller than that of uncoated cBN. (3) The wear progress of the (Ti,Al)N-coated cBN with the main element phase of the TiCN-Al, was slower than that of the (Ti,Al)N-coated cBN with the main element phase of the TiN-Al. (4) In the case of the high cutting speed of 5.00 m/s, the tool wear of the (Ti,Al)N-coated cBN was also smaller than that of uncoated cBN. The above results clarify that the (Ti,Al)N-coated cBN can be used as a tool material in high feed cutting of hardened steel.

1. Introduction

Many difficult-to-cut materials such as hardened steels, titanium and nickel based alloys, etc., are widely used today. For dimensional accuracy, these difficult-to-cut materials are required to be machined by the metal removal process. As these difficult-to-cut materials are required to be machined under high efficiency to improve productivity, it is necessary that the tool materials have good wear-resistance. Polycrystalline cubic boron nitride compact (cBN) seems to be an effective tool material because it has better features as a tool material such as good heat resistance, high hardness, etc. Therefore, regarding the cutting of hardened steel, which is one of these difficult-to-cut materials, there are many studies on the tool wear [1-3], the cutting force [4]-[5] or the surface roughness [6] of cBN tools. The cutting performance of cBN tools depends on the cBN contents rate [7], and the binding phase [8]. Therefore, an effective binding phase, etc. for cBN tools should be selected for the cutting of hardened steel.

On the other hand, coating has long been applied to high speed steel and carbide tools, and more recently, on cBN substrates [9]. Reginaldo T. Coelho et al. reported that in turning a hardened ASTM A29 4340 nickel chromium molybdenum steel (AISI 4340) using three types of coated cBN tool, namely the (Ti,Al)N-nano-coated, (Ti,Al)N-coated and (Al,Cr)N-coated, and uncoated cBN tool, the

lowest tool wear happened with the (Ti,Al)N-nano-coated cBN tool followed by the (Ti,Al)N-coated, (Al,Cr)N-coated and uncoated cBN tool [9].
A hardened ASTM 52100 bearing steel (DIN 100Cr6, 62 HRC) was turned by two types of coated cBN tool, namely the (Ti,Al)N- and TiN-coated cBN tool, and uncoated cBN tool. As a result, the experimental results showed that in the case of the lower feed rate of 0.08 mm/rev, there is little difference in tool life between the coated and uncoated cBN. However, in the case of the higher feed rate of 0.15 mm/rev, the tool life of coated cBN is longer than that of the uncoated tool, in particular the tool life of (Ti,Al)N-coated cBN is three times longer than that of the uncoated cBN tool [10].
It was reported that the coated cBN tool demonstrated longer tool life (~20%) than the uncoated tool life at a cutting speed of V<300 m/min, and the difference in tool life between the coated and uncoated CBN tool is negligible when V>300 m/min in high speed turning of Inconel 718 with the uncoated and coated cBN tool [11]. Moreover, it was reported that at a cutting speed of 250 m/min, the coated cBN tools had approximately 20% longer tool life than uncoated cBN, and the gap was rapidly closing with the increase in speed and became negligible at a speed of 350 m/min in turning of Inconel 718 with the uncoated and coated cBN tools [12].
In high turning Inconel 718 with the coated and uncoated cBN tool at a cutting speed of 250 m/min, the coated cBN tool had approximately 20% longer tool life than the uncoated cBN tool [13].
However, R. M'Saoubi et al. reported that the uncoated insert displays a lower flank wear and crater wear width when compared to most of the coated tools, except TiSiN, excepted in cutting hardened steel 16MnCr5 (58-62 HRC) with four typeskinds of coated cBN tools [14]. Furthermore, Fatih Taylan et al. reported that in face milling of hardened 90MnCrV8 tool steel (61 HRC), the tool failure was also investigated [15]. As a resultresults, the damage of the cutting tools over the entire range of cutting conditions was mainly in the form of chipping, the breakage ISO type C (ISO 8688 standard) of the TiN-based-coated cBN was four times greater than that of uncoated cBN. Therefore, TiN-based-coated cBN was not useful for hard milling application.
In this study, in order to verify the effectiveness of the (Ti,Al)N-coated cBN, which is formed on the substrate of cBN by the physical vapor deposition (PVD) method, the hardened steel was turned with the (Ti,Al)N-coated cBN tool at a cutting speed of 3.33, 5.00 m/s, a feed rate of 0.3 mm/rev and a depth of cut of 0.1 mm. Furthermore, the uncoated cBN, which was the substrate of the (Ti,Al)N-coated, was also used.
The hardened steel used was an ASTM D2 cold-worked die steel (60 HRC). The substrate of (Ti,Al)N-coated cBN used was three types of cBN tools, which had a different contents rate and different main binding phases. The tool wear of the cBN tools was experimentally investigated.
This investigation will contribute to the improvement of productivity in the case of high feed cutting hardened steels.

2. Experimental procedure

The work material used is hardened ASTM D2 cold-worked die steel (60 HRC). The chemical composition of the hardened steel is shown in Table 1.
The tool materials used were a (Ti,Al)N-coated cBN by the PVD method. The substrate of the (Ti,Al)N-coated cBN is shown in Table 2. Furthermore, an uncoated cBN, which was the substrate of the (Ti,Al)N-coated cBN, was also used. The main element phase of the cBN A, cBN B and cBN C is TiCN-Al, TiN-Al and TiN-Al, respectively. The three types of cBN have different cBN content rates and different cBN particle sizes as shown in Table 2. Therefore, the three types of cBN have different mechanical properties, namely hardness and the transverse-rupture strength.
The configuration of the tool insert was ISO TNGA160408. The insert was attached to the tool holder MTGNR2525M16. In this case, the tool geometry was (-6, -6, 6, 6, 29, -1, 0.8 mm). Table 3 shows the cutting conditions.
Figure 1 shows a chamfered and honed cutting edge. In this figure, a chamfer width "W" of 0.130 mm, a chamfer angle "θ" of -25° and a radius of the honed edge "R" of 0.02 mm were employed.

The turning tests were conducted on a precision lathe (Type ST5, SHOUN MACHINE TOOL Co., Ltd.) by adding a variable-speed drive. The driving power of this lathe is 7.5/11 kW with a maximum rotational speed of 2500 min^{-1}. Hardened steel was turned under the cutting conditions shown in Table 3, and the tool wear was investigated.

Table 1. Chemical composition of the hardened steel (AISI D2, 60HRC).

[mass%]

C	Cr	Mo	Mn	Si	V
1.47	11.5	0.82	0.37	0.32	0.20

Table 2. Properties of polycrystalline cubic boron nitride compact.

Tool material	Binding phase *	Contents rate **	cBN particle size [μm]	Hardness [HV]	T. R. S. [GPa]
cBN A	TiCN-Al	45/55	5.0	2700-2900	0.80-0.90
cBN B	TiN-Al	65/35	3.0	3200-3400	1.00-1.10
cBN C	TiN-Al	75/25	5.0	3500-3700	1.15-1.30

*: Main element, **: (cBN grain/ binding phase), T. R. S.: Transverse-rupture strength

Table 3. Cutting conditions.

Cutting speed	V=3.33, 5.00 m/s
Feed rate	f=0.3 mm/rev
Depth of cut	ap=0.1 mm
Cutting method	Dry

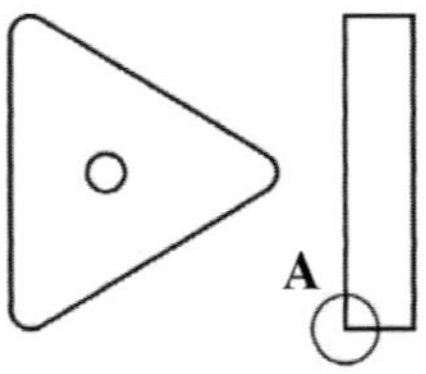

(a) cBN turning insert

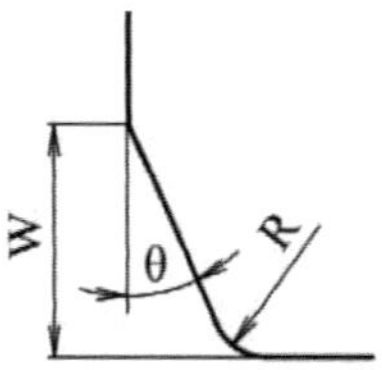

(b) Detail of A

Figure 1. Configuration of a chamfered and honed cutting edge.

3. Results and discussion

Figure 2 shows the tool wear of uncoated and (Ti,Al)N-coated cBN tools. Figures (i), (ii) and (iii) show the case of cBN A, cBN B and cBN C, respectively. Figures (a) and (b) show the case of the uncoated cBN and (Ti,Al)N-coated cBN, respectively. In this figure, "L" is the cutting distance. In all the cBN tools, the grooved abrasive traces can be found on the worn surface of the flank, and the flank wear is parallel to the cutting edge.

Therefore, the maximum value of the flank wear width (VBmax) was measured with a microscope.

Figure 3 shows the relation between the cutting distance and the flank wear width at a cutting speed of 3.33 m/s with uncoated cBN A and (Ti,Al)N-coated cBN A tools. The wear progress of the (Ti,Al)N-coated cBN A is slower than that of uncoated cBN A.

As described above, the wear progress of the (Ti,Al)N-coated cBN A is slower than that of uncoated cBN A. The reason for this is as follows. At first, Figure 4 shows the detail of the worn surface on the rake face. Namely, Fig. 4(a) shows the uncoated cBN A and the detail of A is shown in Fig. 2 (i)(a), and Fig. 4(b) shows the (Ti,Al)N-coated cBN A and the detail of B is shown in Fig. 2 (i)(b). In Fig.

4(b), the "A" shows the contact width between the rake face and the chip and "B" shows the contact length between the rake face and the chip.

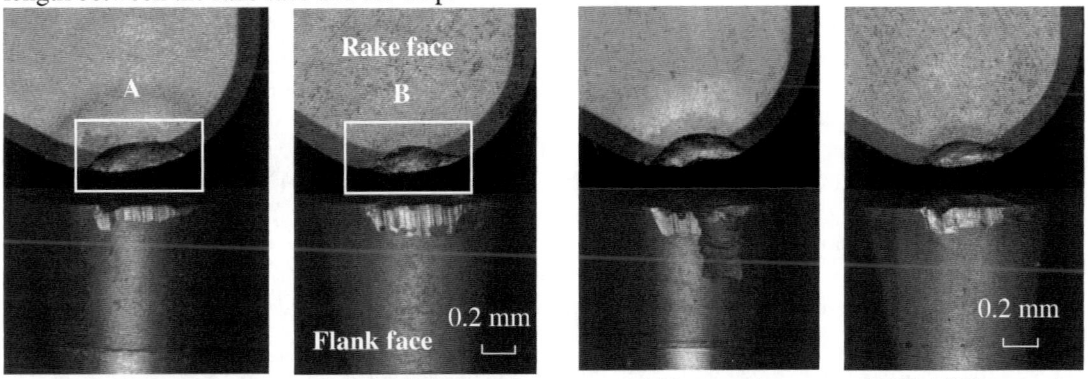

(a)Uncoated, L=225 m (b)(Ti,Al)N-coated, L=445 m
(i) cBN A

(a) Uncoated, L=372 m (b)(Ti,Al)N-coated, L=372 m
(ii) cBN B

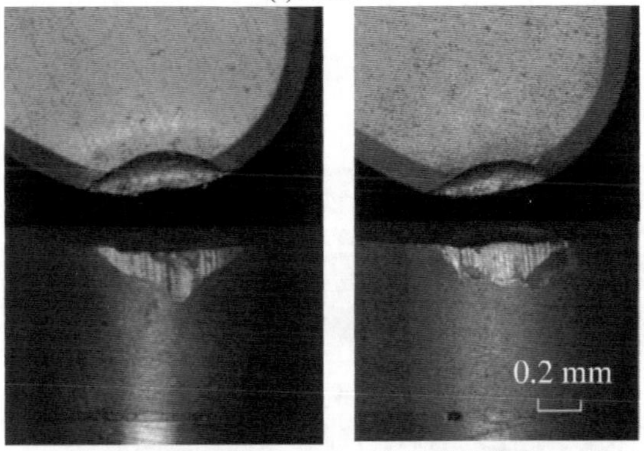

(a) Uncoated, L=225 m (b)(Ti,Al)N-coated, L=298 m
(iii) cBN C

(L: Cutting distance)

Figure 2. Tool wear in cutting hardened steel at a cutting speed of 3.33 m/s, feed rate of 0.3 mm/rev and depth of cut of 0.1 mm.

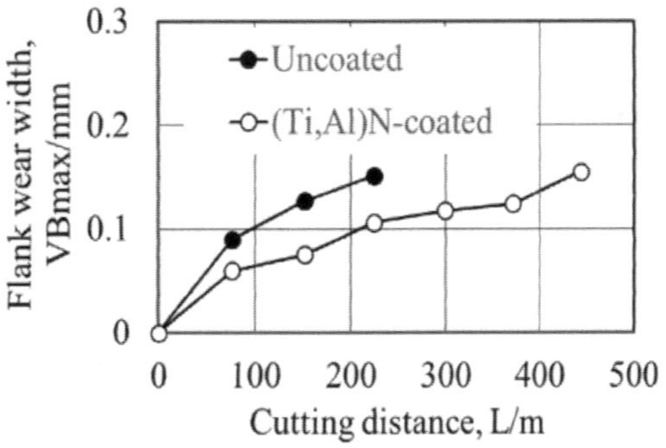

Figure 3. Relation between the cutting distance and flank wear width at a cutting speed of 3.33 m/s with (Ti,Al)N-coated cBN A and uncoated cBN A.

Figure 4 shows the wear photographs on the rake face of cBN A. Table 4 shows the "S" and the dimensions of "A" and "B" shown in Fig. 4. In Table 4, the "S" is the area of the worn surface on the rake face shown in Fig. 4. Compared with uncoated and (Ti,Al)N-coated cBN tool, both the "S" and

the "A", "B" of the (Ti.Al)N-coated cBN tool are smaller than that of the uncoated cBN tool in the case of all cBN tools. The contact area between the rake face and the chip of the (Ti.Al)N-coated cBN tool is smaller than that of the uncoated cBN tool. That is, the (Ti,Al)N-coated cBN A has a smaller contact area between the rake face and the chip as compared with the uncoated cBN A.
Therefore, it is considered that in the case of cutting hardened steel by the (Ti,Al)N-coated cBN A tool, the cutting temperature decreases with the decrease of the friction force between the rake face and the chip. R. M'Saoubi et al. reported also that the tool temperature of (Ti,Al)N-coated cBN is lower, about 10 °C, than that of uncoated cBN in cutting hardened steel 16MnCr5 (58-62 HRC) with the four types of coated cBN tools [14]. This is considered to be the reason why there is less tool wear of the (Ti,Al)N-coated cBN A tool.
Next, Figure 5 shows the EDS mapping analysis on the worn surface of the two types of uncoated and (Ti,Al)N-coated cBN A tools. In Fig. 5, the EDS analysis for the iron (Fe) and oxygen (O) mapping on the cutting part is shown. As compared with the iron element on the worn surface of the uncoated cBN A tool and that of the (Ti,Al)N-coated cBN A tool, the Fe element on the uncoated cBN A tool shown in Fig. 5(a) is less than that on the (Ti,Al)N-coated cBN A tool shown in Fig. 5(b) due to the cutting distance of the (Ti,Al)N-coated cBN A tool that is almost twice that of the uncoated cBN A tool. However, there is little difference between the oxygen element of the uncoated cBN A and that of the (Ti,Al)N-coated cBN A tool, which was turned at a long cutting distance. Therefore, the cutting temperature of the (Ti,Al)N-coated cBN A tool is about the same as that of the uncoated cBN A tool despite the increase in the cutting distance of the (Ti,Al)N-coated cBN A tool.

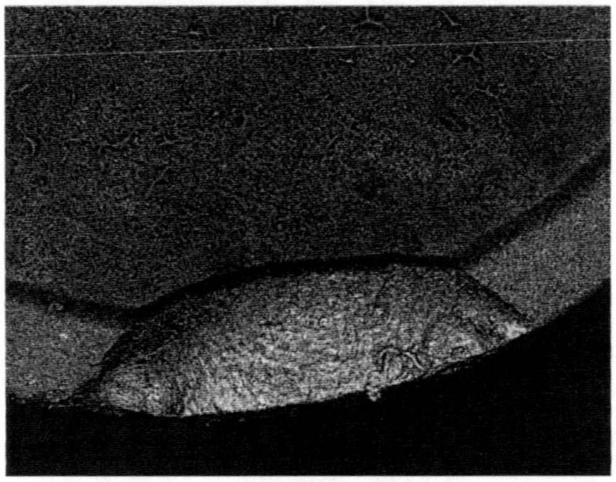

(a) Uncoated
(Detail of A shown in Fig. 2 (i)(a))

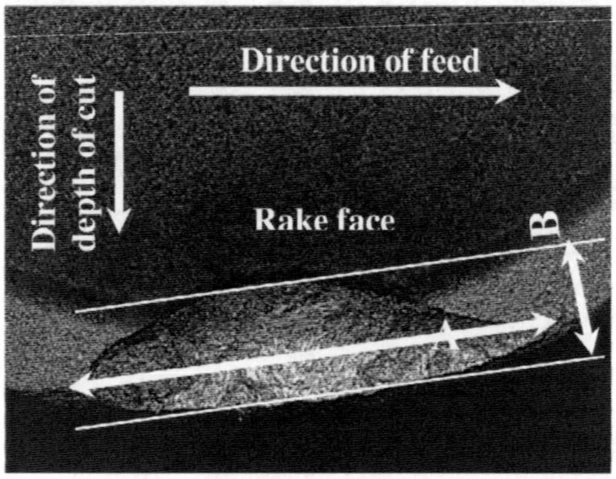

(b) (Ti,Al)N-coated
(Detail of B shown in Fig. 2(i)(b))

Figure 4. Wear photographs on the rake face of cBN A.

Table 4. "S", "A" and "B" shown in Fig. 4.

Tool material	Coating film	**S [x10^{-3} mm^2]**	**A [mm]**	**B [mm]**
cBN A	Uncoated	68.9	0.575	0.173
	(Ti,Al)N	53.7	0.565	0.140
cBN B	Uncoated	80.0	0.574	0.220
	(Ti,Al)N	63.8	0.498	0.161
cBN C	Uncoated	91.3	0.642	0.211
	(Ti,Al)N	64.2	0.588	0.166

S: Area of worn surface on the rake face shown in Fig. 4
A: Contact width between the rake face and the chip shown in Fig. 4(b)
B: Contact length between the rake face and the chip shown in Fig. 4(b)

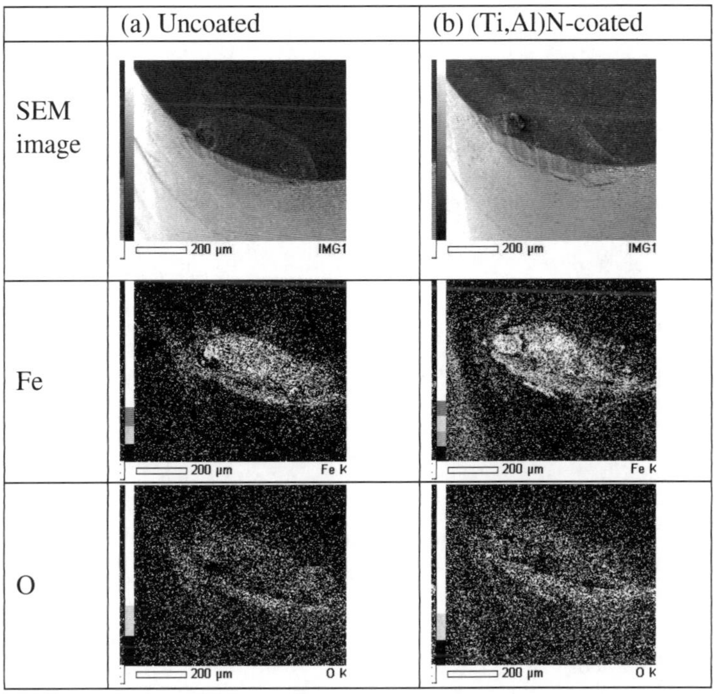

Figure 5. EDS mapping analysis of iron and oxygen on the worn surface (a) uncoated, (b) (Ti,Al)N-coated cBN A tool.

Figure 6 shows the relation between the cutting distance and the flank wear width at a cutting speed of 3.33 m/s with the three types of cBN tools. The wear progress of the (Ti,Al)N-coated cBN A is slowest. Comparing the cBN particle size of the three types of the cBN shown in Table 2, the three types of cBN have approximately the same cBN particle size of 3.0- 5.0 μm. Next, comparing the cBN contents rate of the three types of the cBN, the cBN contents rate of the cBN A, which has 45% cBN content, is the least. Furthermore, both the hardness and the transverse rupture strength of cBN A are the least, too. However, the tool wear of the (Ti,Al)N-coated cBN A is the smallest among the three types of cBN tool. This is due to the following reasons. That is, in the case of high feed rate cutting, the cutting temperature rises. K. Shintani et al. carried out a heated test on cBN tools with various binding phases under the experimental conditions of a temperature of 1273 K and a holding time of 600 s in an Ar atmosphere, and it was inferred that because a part of TiN, which is the main binding phase of cBN, changes to the brittle material TiO_2, the binding strength of the cBN decreases [16]. Therefore, at high feed rate cutting, because a part of TiN, which is the main binding phase of both the cBN B and the cBN C tool, changes to the brittle material TiO_2, the binding strength of the cBN decreases, and the tool wear of both the cBN B and the cBN C becomes larger than that of the cBN A, which has the main binding phase of TiCN-Al.

As described above, the (Ti,Al)N-coated cBN A tool is the most effective tool material in the cutting of hardened steel.

Figure 7 shows the relation between the cutting distance and the flank wear width at a cutting speed of 5.00 m/s with two types of cBN A tools. In the case of the higher cutting speed, the wear progress of the (Ti,Al)N-coated cBN A is also slower than that of the uncoated cBN A.

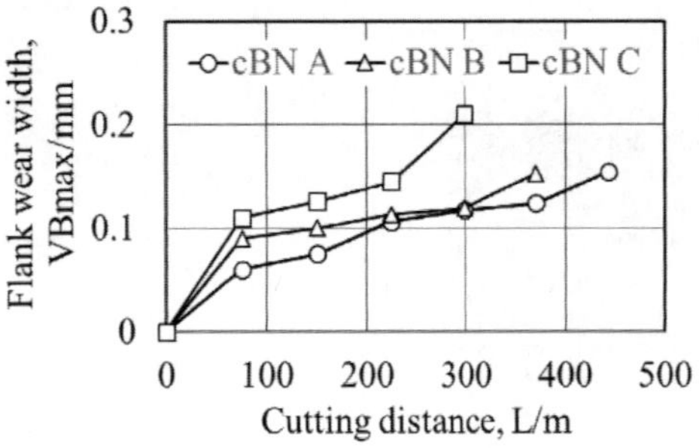

Figure 6. Relation between cutting distance and flank wear width at a cutting speed of 3.33 m/s with the three types of (Ti,Al)N-coated cBN tools.

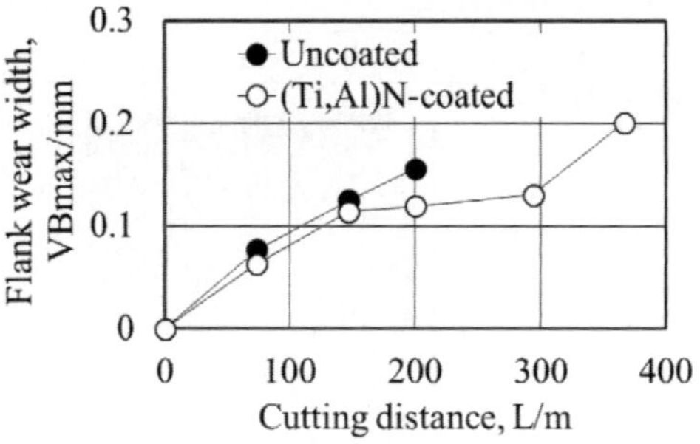

Figure 7. Relation between cutting distance and flank wear width at a cutting speed of 5.00 m/s with (Ti,Al)N-coated cBN A and uncoated cBN A.

4. Conclusion

In this study, in order to verify the effectiveness of the (Ti,Al)N-coated cBN, which is formed on the substrate of cBN by the PVD method, the hardened steel was turned with the (Ti,Al)N-coated cBN tool at a cutting speed of 3.33, 5.00 m/s, a feed rate of 0.3 mm/rev and a depth of cut of 0.1 mm. Furthermore, the uncoated cBN, which was the substrate of the (Ti,Al)N-coated, was also used. The tool wear of the cBN tools was experimentally investigated.

The following results were obtained:

- The contact area between the rake face and the chip of the (Ti.Al)N-coated cBN tool was smaller than that of the uncoated cBN tool.
- The tool wear of the (Ti,Al)N-coated cBN was smaller than that of the uncoated cBN.
- The wear progress of the (Ti,Al)N-coated cBN with the main element phase of the TiCN-Al was slower than that of the (Ti,Al)N-coated cBN with the main element phase of the TiN-Al.
- In the case of the high cutting speed of 5.00 m/s, the tool wear of the (Ti,Al)N-coated cBN was also smaller than that of the uncoated cBN.

These results clarify that the (Ti,Al)N-coated cBN can be used as a tool material in high feed cutting hardened steel.

5. References

[1] Z.N. Farhat, Wear mechanism of CBN cutting tool during high-speed machining of mold steel. *Materials Science and Engineering*. **A361** (2003) pp 100-110.

[2] Gérard Poulachon, B.P Bandyopadhyay, I.S Jawahir, Sébastien Pheulpin, Emmanuel Seguin, Wear behavior of CBN tools while turning various hardened steels. *Wear*. **256, 3-4** (2004) pp 302-310.

[3] Y. Sahin, Comparison of tool life between ceramic and cubic boron nitride (CBN) cutting tools when machining hardened steels. *Journal of Materials Processing Technology*. **209-7** (2009) pp 3478-3489.

[4] Li Qian, Mohammad Robiul Hossan, Effect on cutting force in turning hardened tool steels with cubic boron nitride inserts. *Journal of Materials Processing Technology*. **191, 1-3** (2007) pp 274-278.

[5] Mohamed Athmane Yallese, Kamel Chaoui, Nassereddine Zeghib, Lakhdar Boulanouar, Jean-François Rigal, Hard machining of hardened bearing steel using cubic boron nitride tool. *Journal of Materials Processing Technology*. **209-2** (2009) pp 1092-1104.

[6] R.M Arunachalam, M.A Mannan, A.C Spowage, Residual stress and surface roughness when facing age hardened Inconel 718 with CBN and ceramic cutting tools. *International Journal of Machine Tools and Manufacture*. **44-9** (2004) pp 879-887.

[7] Zhengwen Pu, Anshul Singh, High speed ball nose end milling of hardened AISI A2 tool steel with PCBN and coated carbide tools. *Journal of Manufacturing Processes*. **15** (2013) pp 467-473.

[8] Tadahiro Wada, Tool Wear of Polycrystalline Cubic Boron Nitride Compact Tools in Cutting Hardened Steel. *Advanced Materials Research*. **488-489** (2012) pp 724-728.

[9] Reginaldo T. Coelho, Eu-Gene Ng, M.A. Elbestawi. *International Journal of Machine Tools and Manufacture*. **47** (2007) pp 263-272.

[10] Gustavo de Siqueira Galoppi, Marco Stipkovic Filho, Gilmar Ferreira Batalha, Hard turning of tempered DIN 100Cr6 steel with coated and no coated CBN inserts. *Journal of Materials Processing Technology*. **179** (2006) pp 146-153.

[11] Jinming Zhou, Volodymyr Bushlya, Pajazit Avdovic, Jan Eric Ståhl, Study of surface quality inhigh speed turning of Inconel 718 with uncoated and coated CBN tools. *International Journal of Advanced Manufacturing Technology*. **58** (2012) pp 141-151.

[12] Volodymyr Bushlya, Jinming Zhou, Pajazit Avdovic, Jan-Eric Ståhl, Performance and wear mechanisms of whisker-reinforced alumina, coated and uncoated PCBN tools when high-speed turning aged Inconel 718. *International Journal of Advanced Manufacturing Technology*. **66** (2013) pp 2013-2021.

[13] V. Bushlya, J. Zhou, J.E. Ståhl, Effect of Cutting Conditions on Machinability of Superalloy Inconel 718 During High Speed Turning with Coated and Uncoated PCBN Tools. *Procedia CIRP*. **3** (2012) pp 370-375 (45th CIRP Conference on Manufacturing Systems 2012).

[14] R. M'Saoubi, M.P. Johansson, J.M. Andersson, Wear mechanisms of PVD-coated PCBN cutting tools. *Wear*. **302** (2013) pp 1219-1229.

[15] Fatih Taylan, Oğuz Çolak, Mehmet Cengiz Kayacan, Investigation of TiN Coated CBN and CBN Cutting Tool Performance in Hard Milling. Application Journal of Mechanical Engineering. **57-5** (2011) pp 417-424.

[16] K. Shintani, H. Katoh, et al., Cutting Performance of CBN Tools in Machining of Austempered Bainitic Spheroidal Graphite Cast Iron. *Journal of Japan Soc. Precision Engineering*. **56-12** (1990) pp 2261-2266. (in Japanese)

Acknowledgment

We would also like to express our sincere thanks to the Tungaloy Corporation for their cooperation in the supply of cBN inserts.

IOP Conf. Series: Materials Science and Engineering **264** (2017) 012018 doi:10.1088/1757-899X/264/1/012018

The effects of foaming conditions on plasticized polyvinyl chloride foam morphology by using liquid carbon dioxide

N Chuaponpat[1] and S Areerat[2]

1 Department of Chemical Engineering, Faculty of Engineering
2 King Mongkut's Institute of Technology Ladkrabang, Bangkok 10520, Thailand
E-mail: surat.ar@kmitl.ac.th

Abstract. This research studies the effects of foaming conditions by using liquid carbon dioxide (CO_2) as a physical blowing agent on plasticized polyvinyl chloride (PVC) foam morphology. Foaming conditions were soaking time of 6, 10, and 12 h, foaming temperature of 70, 80, 90 ^{o}C for 5 s, at constant soaking temperature of -20 ^{o}C and pressure of 50 bar. Instantaneously increasing temperature was employed in this process for making foam structure. PVC foam samples were calculated percentage of shrinkage (Sh) by using density at before and after aging process at 30 ^{o}C for 12 h. When PVC samples were activated to form foam by using liquid CO_2 as a physical blowing agent, it reveal bimodal foam structure with a thick bubble wall (10-20 μm). Bubble diameter of PVC foam at longer soaking time is in the range of 40-60 μm and its at shorter soaking time reveal a large bubble that is in the range of 80-120 μm. Foaming condition slightly affected to bubble density that was in the narrow range of 10^6-10^8 bubbles/cm^3. PVC foam reveal reduction of density up to 65% when compare with PVC and Sh is less than 10%.

1. Introduction

Polymer foam is defined as gas dispersed in polymer matrix such as polystyrene foam, polyethylene foam, polypropylene foam, polyvinyl chloride foam etc. Dispersed gas in polymer matrix can be generate into open-cell and close-cell structure [1]. The advantages of foam property are strength-to-weight ratio, low thermal conductivity, low electric constant, low density when compared with solid material. Thus, there are many applications of polymer foam such as auto motive parts, toys, electrical wire, food packaging, etc [2]. Each application require differential foam property, which relate to foam morphology and aging process after foam formation [3]. Therefore, there are three parameter to control foam morphology, polymer property, foaming process, and additive. Polymer properties are molecular weight, molecular weight distribution, melt flow index, polymer chain structure, amorphous polymer, semi-crystalline polymer, molecular orientation etc. When polymer property is optimized, foaming process is desired from polymer property such as, extrusion, injection mold, compression etc. Foaming process also consist of foaming conditions, temperature, time, pressure, flow rate, depressurization rate etc [4-7]. However, additive is chemical and/or physical material that is employed to adjust foam morphology such as, blowing agent, nucleating agent, plasticizer, fire retarding agent, etc. All of three parameters should be optimized to control foam morphology.

Basically of foaming process with using gas as physical blowing agent, there are four steps of foaming formation as shown in Figure 1 [8]. First, gas diffuse and dissolve in polymer matrix at desired conditions, time, pressure and temperature into heterogeneous/heterogeneous solution. Then, the system is activated by increasing temperature or decreasing pressure and nucleus generate during nucleation step. After that, nucleus expand into a bubble or cell form during cell growth and then finally step is cell stabilization by air permeation into foam structure to replace gas.

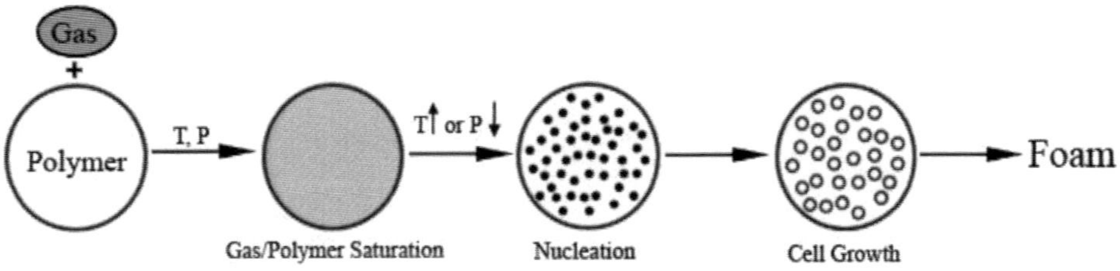

Figure 1. Foam formation for using gas as physical blowing agent.

This research selected amorphous polymer or plasticized polyvinyl chloride (PVC) to study PVC foam morphology because it is high margin of the construction industry in Thailand. PVC is foamed by using liquid carbon dioxide (CO2) as blowing agent then it was left for aging process to observe shrinkage property.

2. Experiment and characterization

2.1 Foaming process

PVC sheet samples were soaked by using liquid CO_2 for 6, 10, and 12 h at constant temperature of -20 °C and pressure of 50 bar, as shown in Figure 2. After soaking step, these samples were immediately taken out and dipped in the silicone oil bath at desired foaming temperature (70, 80 and 90 °C) for 5 s to activate nucleation. PVC samples were cleaned and left for stabilization at 30 °C for 12 h then these samples were calculated the percentage of shrinkage (Sh) of the sample. The percentage of shrinkage is defined as the changing volume of PVC foam sample between before and after aging process and calculated by equation (1) which based on mass constant and neglect gaseous mass.

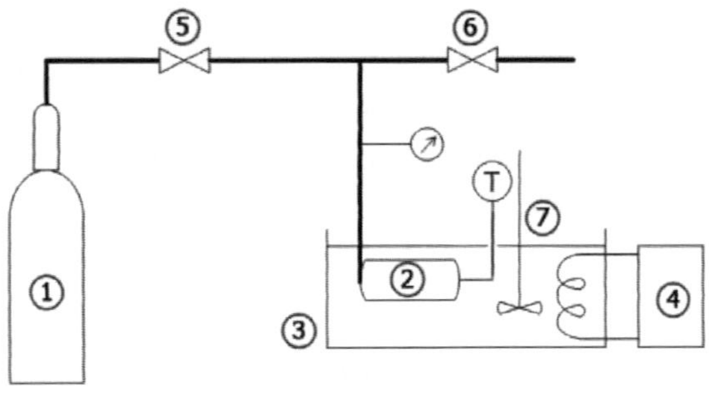

Figure 2. Schematic of soaking process by liquid CO_2 which including of CO_2 tank (1), 18 ml high pressure vessel (2), cooling bath (3), cooling unit (4), ball valves (5,6), and stirrer (7).

$$Sh = \left(\frac{V_{be}-V_{af}}{V_{be}}\right) \times 100 = \left(1 - \frac{\rho_{be}}{\rho_{af}}\right) \times 100 \quad (1)$$

when

V_{be}, V_{af}: before and after aging volume of PVC foam specimens, respectively (cm^3)

ρ_{be}, ρ_{af} : before and after aging density of PVC foam specimens, respectively (g/cm^3)

2.2 Foam morphology characterization

PVC foam samples were dipped in the liquid nitrogen (purchased from Linde (Thailand) Public Limited Company) for 30 s and cracked to observe PVC foam morphology at cross-sectional area.

IOP Conf. Series: Materials Science and Engineering **264** (2017) 012018 doi:10.1088/1757-899X/264/1/012018

These samples were coated by using the SC7620 sputter coater (Quorum Technologies Ltd.) and then were observed the foam morphology by using scanning electron microscope (MA10 from Carl Zeiss). SEM micrographs were characterized to obtain bubble diameter and bubble density (N), which calculate by equation (2) [9] as a function of number of bubbles (n) in observed cross-sectional area (A).

$$N = \left(\frac{n}{A}\right)^{3/2} \tag{2}$$

3. Results and discussion

PVC samples were soaked by liquid CO_2 at -20 °C, which below glass transition temperature of its, then it was activated foam by rapidly increasing temperature at desired foaming temperature. PVC density of before and after aging samples at various foaming conditions was presented in Figure 3, which can be represented the effect of solubility on foam density. Before aging PVC density at constant foaming temperature of 70 °C slightly decrease from 0.454 g/cm^3 to 0.399 g/cm^3 as soaking time but it strongly reduce from 0.695 g/cm^3 to 0.556 g/cm^3 at constant foaming temperature of 90 °C because liquid CO_2 molecules have more time to dissolve in PVC matrix and high nucleation rate at high temperature. After aging process, PVC foam density increased about 30%, it might be caused by rapidly diffusion rate of CO_2 escape from PVC structure that structure collapsed. Thus, PVC samples reveal percentage of shrinkage lower than 10%, as shown in Figure 4.

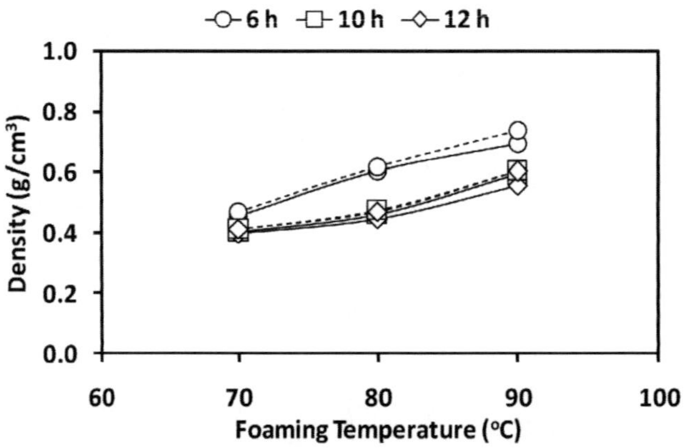

Figure 3. Before (solid line) and after (dashed line) aging density of PVC samples.

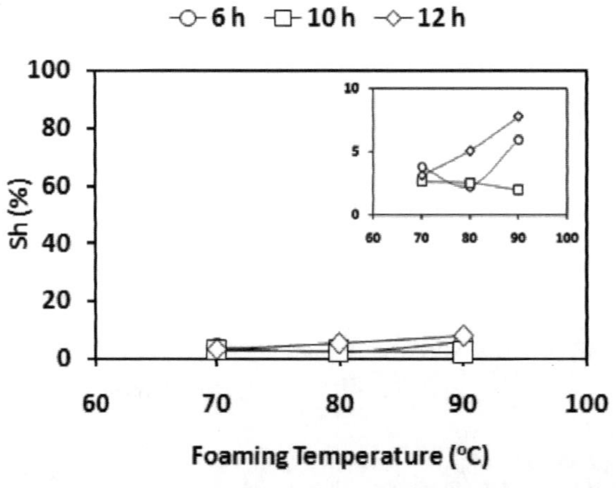

Figure 4. Percentage of shrinkage of PVC samples.

PVC samples morphology were observed by using SEM and shown in Figure 5. Effect of soaking time can be observed from top to bottom and effect of foaming temperature can be analyzed from left to right. PVC samples morphology have non-uniform and bimodal foam structure. At constant soaking time of 6 h, PVC samples morphology contain closed and slight open bubble structure with thick bubble wall. It might be caused by short time to dissolve of liquid CO_2 in PVC matrix that high melt strength restrict to bubble formation while nucleated bubble collapse and fuse with neighbour bubble. Thus, it is the highest foam density than other foaming conditions. Increasing foaming temperature affect to enhance nucleation rate that PVC samples can be observed more bubbles population with micro-scale. Improving nucleation rate can be involved by increasing solubility as soaking time of liquid CO_2 in PVC up to 10 and 12 h. PVC samples morphology reveal higher bubble population and bimodal foam structure, which approximately bubble size is larger than 50 μm and less than 10 μm. So, PVC samples with high solubility exhibit lower foam density than it at low solubility of liquid CO_2. Due to short foaming time during nucleation step, PVC matrix was not fully soft matrix, thus PVC samples display bimodal foam structure with thick bubble wall. That affect to high foam density and low shrinkage property.

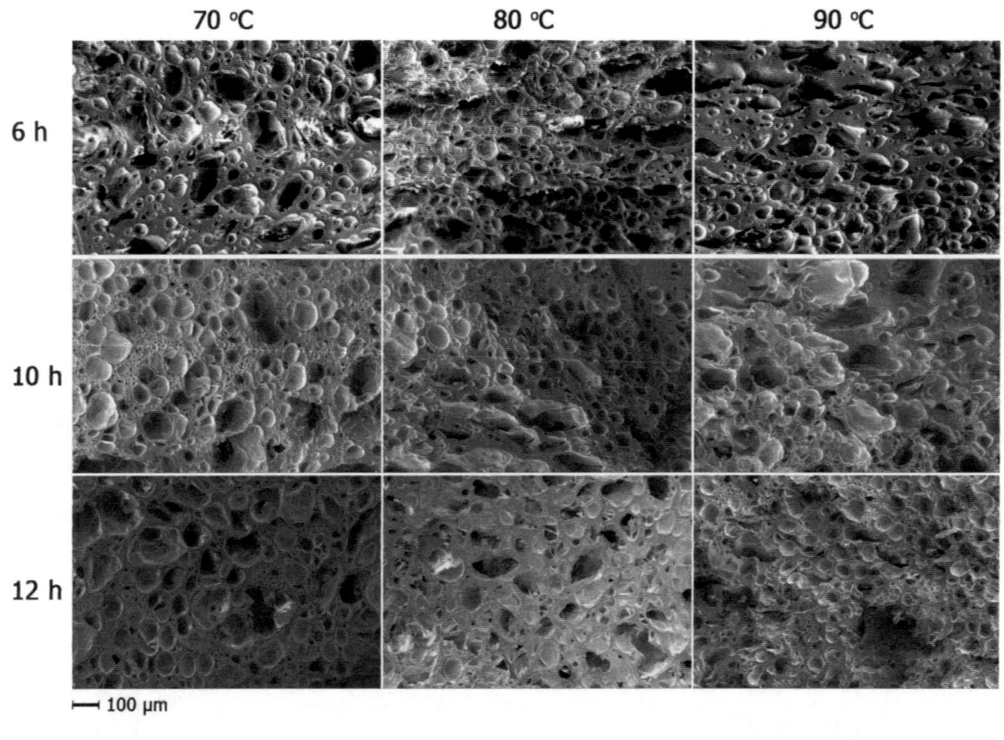

Figure 5. SEM micrographs of PVC samples as various foaming conditions.

As previous results, bubble diameter and bubble density of PVC samples were measured and plotted as a foaming condition, which shown in Figure 6 and Figure 7, respectively. PVC samples at constant soaking time of 6 h reveal 110 μm of bubble diameter at foaming temperature of 70 °C then bubble diameter drastically reduce to 75 μm and 70 μm as foaming temperature increased to 80 °C and 90 °C, respectively. It might be caused by enhancing nucleation rate and decreasing melt strength as increasing foaming temperature at low solubility of liquid CO_2 in PVC. However, critical radius is still a large size so nucleus collapse and fuse with neighbour nucleus to a large form or open bubble structure. When soaking time of CO_2/PVC system was increased up to 10 h and 12 h, more CO_2 diffused and dissolved into PVC matrix, which reduced melt strength PVC matrix. So that, nucleus formed and expanded into a bubble within the range of 40-60 μm. Moreover, increasing a foaming temperature from 70 °C to 90 °C at constant soaking time slightly affects to reduce bubble diameter.

To improve that foaming conditions enhance nucleation rate, bubble density was represented in Figure 7. Bubble density of PVC samples at constant soaking time of 6 h significantly increased from $3x10^6$ to $2x10^7$ bubbles/cm^3 due to enhance nucleation rate by increasing foaming temperature. Liquid CO_2 in solid PVC matrix is slow diffusion, so that it favors to take a long time to achieve the saturated condition. Soaking time of CO_2/PVC system was extended to 10 h and 12 h, respectively, so that melt strength also reduced because of the plasticization effect of dissolved CO_2. Hence, PVC samples at soaking time of 10 h and 12 h reveal higher bubble density than PVC samples at soaking time of 6 h. On the other hand, increasing foaming temperature at soaking time of 10 h and 12 h slightly affect to the bubble density of PVC samples.

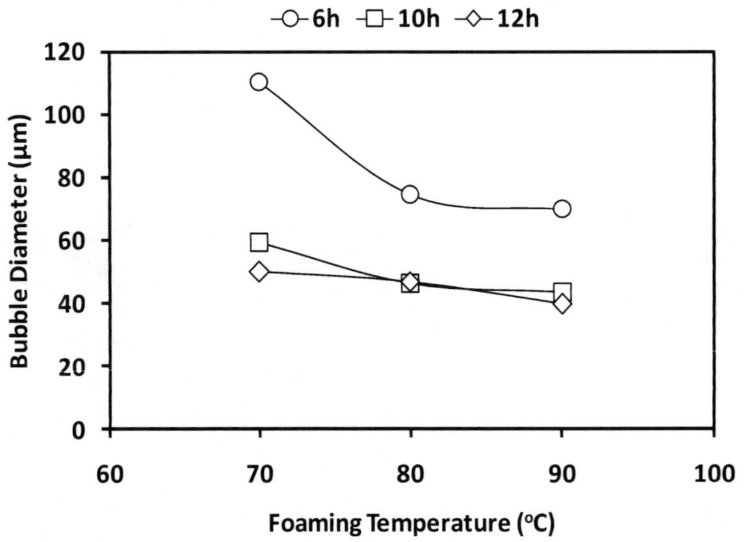

Figure 6. Effect of foaming temperature on bubble diameter of PVC samples.

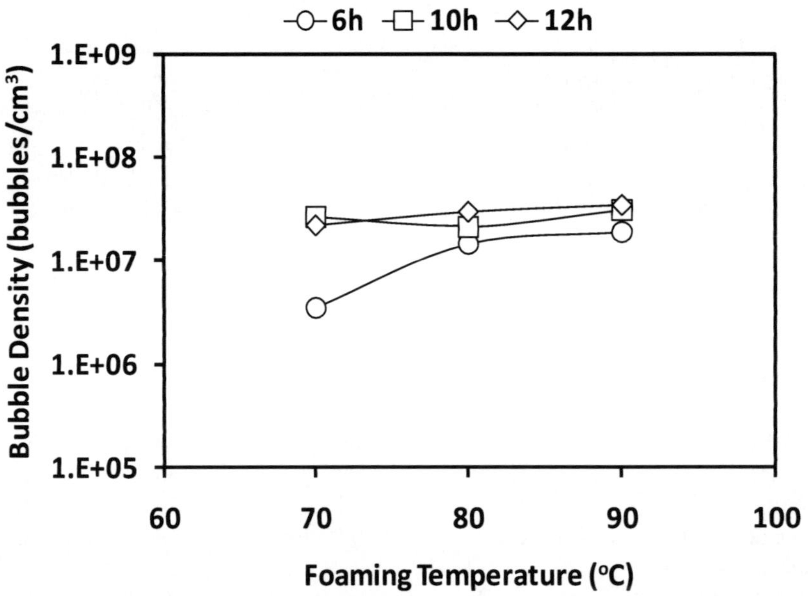

Figure 7. Effect of foaming temperature on bubble density of PVC samples.

4. Conclusion

Liquid CO_2 was employed as physical blowing agent during foaming process. PVC foam samples reveal bimodal foam structure with a thick bubble wall and partly exhibit slight open bubble structure. PVC foam samples density is in the range of 0.4 – 0.6 g/cm^3 and after aging foam density slightly increased as before aging foam density. Thus, PVC foam samples demonstrate 10% of Sh.

5. References

[1] Klempner D and Frisch K C 1991 *Handbook of Polymeric Foams and Foam Technology* (New York: Hanser Publishers)
[2] Tsivintzelis I, Angelopoulou A G and Panayiotou C 2007 *Polymer.* **48** 5928-5939
[3] Lee S T and Ramesh N S 2004 Polymeric Foams: Mechanisms and Materials (New York: CRC Press)
[4] Doroudiani S and Kortschot M T 2003 *J. Appl. Polym. Sci.* **90** 1412– 1420
[5] Taki K 2008 *Chem. Eng. Sci.* **63** 3643 – 3653
[6] Shafi M A, Lee J G and Flumerfelt R. W 1996 *Polym. Eng. Sci.* **36** 1950 – 1959
[7] Liu T, Li D C, Zhao L and Yuan W K 2010 *Particuology.* **8** 607– 612
[8] Xu Z M, Jiang X L, Liu T, Hu G H, Zhao L, Zhu Z N and Yuan W K 2007 *J. Supercrit. Fluid.* **41** 299– 310
[9] Gosselin R and Rodrigue D 2005 *Polym. Test.* **24** 1027-1035

Acknowledgments
The authors are grateful to the Thai Plastics and Chemicals Co., Ltd. for financial and material support. The first author is thankful for Mr. Nantakit Mairungrot, Miss Kanokwan Kankate, and Miss Varintorn Traisattayawej.

In situ viscoelasticity and in situ thermo-responsiveness in acrylic acid-based soft hydrogels

Saminu M. Magami

Material Chemistry and Processing Unit, School of Engineering & the Built Environment, Edinburgh, Napier University, United Kingdom
s.magami@napier.ac.uk

Abstract. This study was aimed at understanding any effects that the presence of acrylic acid dimer could have on thermo-mechanical properties of acrylic acid-based hydrogels. Polymeric hydrogels were systematically synthesised and characterised on the basis of having different concentrations of acrylic acid dimer. The materials show a reversible thermo-responsive behaviour, when heated in situ using light transmittance approaches. The point of cloudiness is significantly influenced as a consequence of the presence of the acrylic acid dimer material. Increasing the amount of the dimer additive up to 2% improves the light transmittance (%) of the soft gels by up to 70%, at 140℃, suggesting that a potential method of altering the lower critical solution temperature of the soft gels is to alter the polymer backbone through copolymerization. As the amount of dimer was increased up to 2%, the values of storage modulus (G') were increased by up to 75%, leading to greater viscoelasticity and stronger gel compositions.

1. Introduction

Hydrogels are polymers that can absorb water, thereby swelling without dissolution [1]. This property arises from the structure of the gels which is based on a three dimensional network, created by physical interactions and chemical cross-linking of the polymer structure [2]. Many hydrogels are created using acrylic acid (AA) as a monomeric component, for various applications [3-5]. Stimuli responsive hydrogel materials, based on hydrophilic polymers such as poly (acrylic acid), have drawn interest from academia and industry, due to their responsive behaviour to external stimuli such as temperature change, changes to the ionic strength and pH [6].

Rheometric, viscoelastic characterisations are often necessary in order to provide understanding of the behaviour of soft gel materials. In such materials, the storage modulus (G') and loss modulus (G") are parameters that are investigated in order to establish the frequency dependency of the gel structure, the stiffness and the energy dissipation events [7]. These are useful in the interpretation of soft gel characteristics such as mechanical rigidity and relaxation. The cloud point and optical properties of hydrogels are influenced by the spatial inhomogeneity in the gel networks. These arise due to a number of factors such as differences in the reactivity ratios of the monomer(s) and cross-linker(s), cyclisation resulting in structural loop formation, microgel formation, diffusion controlled reactions, crosslink density, pH of the composition, ionic strength and the nature of any cations, present as the counter ion [8].

In this study, controlled, small quantities of AA-dimer were used in a range of gel-precursor formulations in order to monitor the contribution of the presence of this AA-dimer to the viscoelastic properties and the thermo-responsiveness of the resulting soft gel materials.

2. Experimental

The gel-precursor solutions were prepared by using methylene bis-acrlylamide (MBA) as crosslinker (0.05%), UV Irgacure IR1170 as photoinitiator (0.01%), acrylic acid as monomer (10%) and water as solvent. The solutions were prepared containing known amounts of the AA-dimer, covering dimer loadings of 0, 0.05, 0.1, 0.25, 0.5, 1.0 and 2.0% of the total formulation. Each solution was stirred, using a magnetic bar, for 5 minutes, then filtered, using 0.2 a micron syringe filter. pH of the aqueous solutions were measured using HI1230 portable pH meter from HANNA Instruments. Ultra-Pure water (conductivity of 0.06 μS/cm and resistivity of 16.6 MΩ) was used for pKa calculations.

In situ photo-crosslinking reactions and rheology were performed using a TA–Instruments DHR2 with photo-curing bottom plate fixture attached to the rheometer. With these, an Omnicure S200S lamp source was used, emitting radiation through a clear glass bottom window. Radiation intensity was fixed at 8 mW/cm2, for 30 minutes cure, while a Smart Swap steel parallel plate fixture with 20 mm diameter was employed using an initial gap size of 2 mm. Frequency sweeps were carried out in the materials' LVR region using a strain value of 0.1%, for the frequency range 20 Hz to 0.02 Hz.

In situ thermo-responsive characteristics were studied using a PerkinElmer Lambda-25 UV-VIS Spectrophotometer, equipped with a cell heating accessory from Pike Technology. Measurements on UV cured gel samples were carried out using 10mm crimp sealed vials designed to withstand temperatures of up to 170 ℃. The transmittance of the hydrogel samples was recorded as a function of temperature by scanning the sample in the visible region of the spectrum (400-700 nm).

3. Results and Discussion

3.1 Effect of AA-dimer on viscoelasticity of the soft gels

Table 1. pKa values for AA and AA-dimer [9].

			pKa values	
Carboxylic Acid	Conc. (M)	pH	This work	Literature
AA	0.14	2.60	4.34	4.23
AA-dimer	0.14	3.05	5.24	4.99

The evolution of the elastic (G') and viscous (G'') moduli for the soft gels, during the photo-polymerisations, are represented in Figure 1. Increasing the amount of AA-dimer in the liquid formulations does not significantly affect the cure patterns. It was also observed that minor differences exist in the cure kinetics. An increase in the AA-dimer loading slightly slows down the initial structure development. Such minor differences can be expected since increasing the AA-dimer loading slightly affects the pH of the liquid formulations, prior to the photo-polymerisation. In addition, it is anticipated that the molar ratio of the photo-initiator/AA monomer is marginally affected by the presence of AA-dimer in the formulations when used as additive, although AA and AA-dimer and have different pKa values, as discussed with respect to Table 1. As Figure 1 and Figure 2 show, the amount of AA-dimer in the formulations affects the mechanical properties of the soft gels which are formed from the photo-polymerisation. This indicates that, by copolymerising AA with AA-dimer, the micro-structure and the mechanical properties of the soft gels can easily be influenced.

According to Figure 1 and Figure 2, those gels with greater amounts of AA-dimer are stiffer than those containing no AA-dimer or smaller amount of AA-dimer. For practical reasons, the stronger the soft gels the better, for their potential application in different areas (biomedical, tissue engineering, et al.). Since G' is significantly greater than G'' value, viscoelastic solid behaviour is clearly demonstrated by the gels. In Figure 2, the elastic moduli were less dependent on frequency as a result of increase in the amount of AA-dimer. This lesser frequency dependence of the modulus indicates greater 'gel' character. The significance of these results is that the viscoelastic properties of soft gels-based on AA can be optimised by using small amounts of AA-dimer as a dopant material.

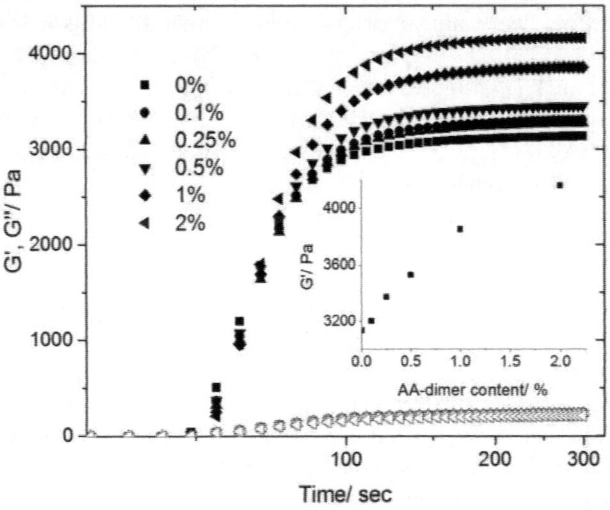

Figure 1. *In situ* cure profile (1Hz, 0.1% strain at RT), showing the evolution of the elastic moduli G' (filled symbols) and loss moduli G'' (open symbol), for compositions containing different amounts of AA-dimer underwent photo-polymerisation

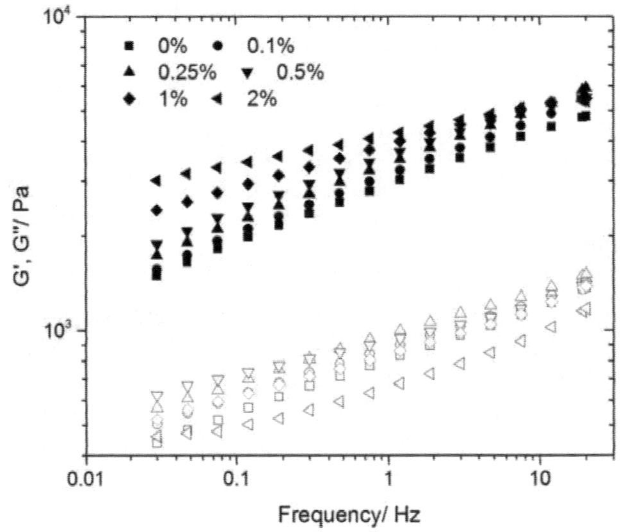

Figure 2. Frequency sweep data showing the elastic moduli G' (filled symbols) and loss moduli G'' (open symbols) as a function of the applied frequency from the soft gels that contain various amounts of AA-dimer. The strain was 0.1% and the run temperature was 25℃

3.2 Effect of AA-dimer on thermo-responsiveness of the soft gels

The effects of temperature change are among the more commonly exploited stimuli in responsive polymeric soft gels [10]. Figure 3 shows a reversible temperature dependent light transmittance. Figure 4 shows the dependence of light transmittance on the amount of AA-dimer used in formulation. It is proposed that the presence or absence of AA-dimer strongly affects the hydrophilic/hydrophobic nature of the polymer, hence the optical transitions (clear/cloudy) are controlled by balance between the hydrophilic and hydrophobic interactions. At elevated temperatures, the carboxyl groups of poly (carboxyethyl acrylate) would remain protonated, leading to a reduction of the hydrophobic complexation by Mg^{2+} ions by maintaining their hydrophilic interactions with water molecules via hydrogen bonding. This phenomenon is illustrated by Figure 5, showing intra-molecular hydrogen bonding and the inter-molecular bonding between the Mg^{2+} and the COO- groups of either the poly(acrylic acid) or the poly(carboxyethyl acrylate) species that are attached to the polymer backbone. Below the critical temperature, poly(acrylic acid) soft gels are clear and transparent suggesting that hydrophilic interactions such as hydrogen bonding between water and the polar functional groups that are present in the matrix, such as –COOH and –C=O, are dominant. However, as the temperature is increased, these hydrophilic interactions are reduces, allowing cations such as Mg^{2+} to complex with the carboxylate ions. This complexation of Mg^{2+} ions causes the composition to become more hydrophobic in nature, leading to precipitation or phase separation, resulting in cloudiness.

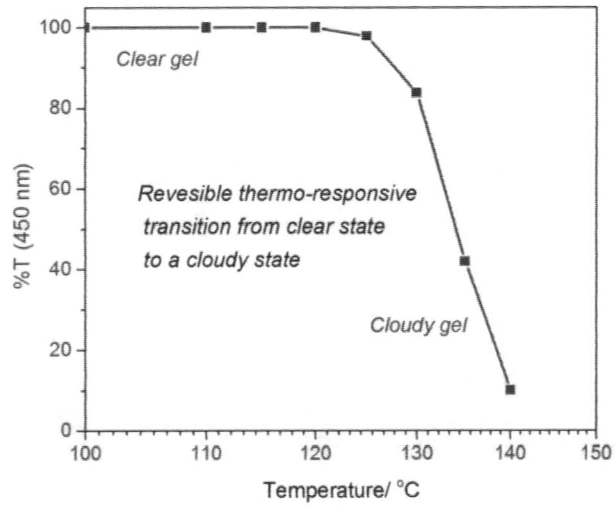

Figure 3. Light transmittance at 450 nm as a function of temperature f for the soft gel that had no AA -dimer added during formulation. Measurements were carried out while the sample was heated in-situ

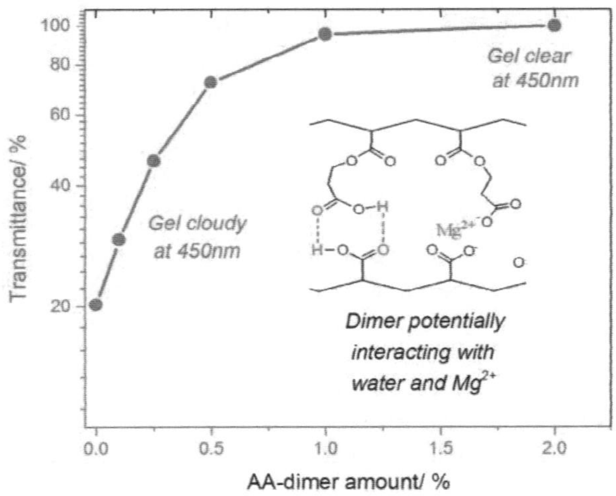

Figure 4. % Light transmittance at 450 nm and cloud point dependence of on AA-dimer content of the soft gels. The light transmittance measurements were carried out at 140℃, the samples being heated in-situ

As the content of AA-dimer was increased, the optical clarity of the soft gel at 140℃ was markedly increased, reaching 100% light transmittance when 2% AA-dimer was present in the polymerisation. The carboxyl groups in the polymer backbone are capable of making strong hydrogen bonds with water molecules both in ionised and protonated form and it may be due to these strong enthalpic interactions that poly(acrylic acid) hydrogels show very high lower critical solution temperature (LCST). The LCST of poly(acrylic acid) hydrogels has previously been shown to be controlled by manipulating a number of parameters such as pH, ionic strength, nature of cation and anion, and by co-polymerising with hydrophobic or hydrophilic (ionic, non-ionic) monomer [11-12].

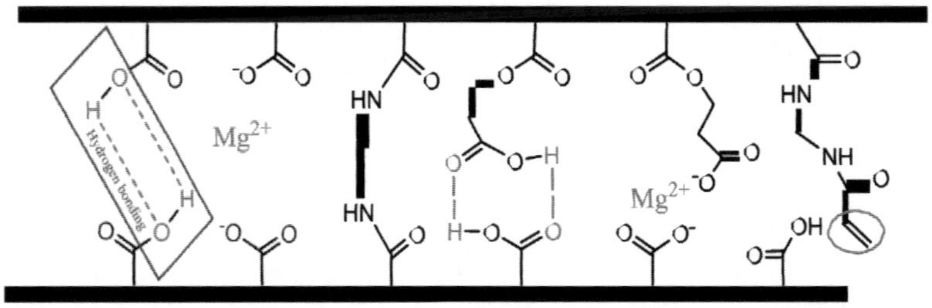

Figure 5. Illustration of the potential interactions, physical and chemical, in soft gels created from AA and containing AA-dimer, with MBA being the crosslinking agent

4. Conclusions

The viscoelastic properties and the thermo-responsive characteristics of soft hydrogels containing AA-dimer were successfully investigated. Both properties were influenced by AA-dimer that was present in the composition. A correlation was observed showing that superior viscoelastic properties of the soft gels could be achieved with an increase in the AA-dimer content. For gel composites containing no or little AA-dimer, a very hazy, semi-opaque gel was observed at 140℃, in contrast to the transparent gel material that was observed having 2% loadings of the AA-dimer in the formulation, at the same temperature.

5. References

[1] K. Kajiwara, Y. Osada, and A. Yamauchi, in: *Polymer Gels - Fundamentals and Biomedical Applications*, vol. 1, Springer publishers, New York (1981).

[2] M. J. Qian, "Synthesis of self-crosslinking sodium polyacrylate hydrogel and water-absorbing mechanism" Journal of Materials Science, 36, 731-738 (2011).

[3] E. Jabbari and S. Nozari, "Swelling behavior of acrylic acid hydrogels prepared by γ-radiation crosslinking of polyacrylic acid in aqueous solution" European Polymer Journal, 36, 2685–2692 (2000).

[4] A. N. Peppas, P. Bures, W. Leobandung and H. Ichikawa, "Hydrogels in pharmaceutical formulations" European Journal of Pharmaceutics and Biopharmaceutics, 50, 27-46 (2000).

[5] T. M. Don, M. L. Huang, A. C. Chiu, K. H. Kuo, W. Y. Chiu and L. H. Chiu, "Preparation of thermo-responsive acrylic hydrogels useful for the application in transdermal drug delivery systems" *Materials Chemistry and Physics*, 107, 266–273 (2008).

[6] R. Debashish, J. N. Cambre and B. S. Sumerlin, "Future perspectives and recent advances in stimuli-responsive materials" *Progress in Polymer Science*, 5, 278–301 (2010).

[7] S. B. Ross-Murphy and G. M. Kavanagh, "Rheological characterisation of polymer gels" *Progress in Polymer Science*, 23, 533–562 (1998).

[8] S. Abdurrahmanoglu and O. Okay, "Homogeneous Poly(acrylamide) Hydrogels Made by Large Size, Flexible Dimethacrylate Cross-Linkers" *Macromolecules*, 41, 7759-7761 (2008).

[9] N. M. Ranjha and U. F. Qureshi, "Preparation And Characterization Of Crosslinked Acrylic Acid/Hydroxypropyl Methyl Cellulose Hydrogels For Drug Delivery" International Journal

of Pharmacy and Pharmaceutical Science, 6, 400-410 (2014).
[10] R. Pelton, "Temperature-sensitive aqueous microgels" *Advance Colloids and Interface Science*, 85, 1–33 (2000).
[11] M. K. Yoo, "Effect of polyelectrolyte on the lower critical solution temperature of poly (N-isopropyl acrylamide) in the poly (NIPAAm-co-acrylic acid) hydrogel" Y. K. Sung, Y. M. Lee and C. S. Cho, *Polymer*, 41, 5713–5719 (2000).
[12] M. S. Jones, "Effect of pH on the lower critical solution temperatures of random copolymers of N-isopropylacrylamide and acrylic acid" European Polymer Journal, 35, 795–801 (1999).

Acknowledgement
This work was supported and funded by the Innovate UK agency, under the Knowledge Transfer Partnership Programme (KTP).

IOP Conf. Series: Materials Science and Engineering **264** (2017) 012020 doi:10.1088/1757-899X/264/1/012020

Room temperature sintering of printer silver nanoparticle conductive ink

Dianne C. Corsino[1] and Mary Donnabelle L. Balela[1*]

[1]Sustainable Electronic Materials Group, Department of Mining, Metallurgical and Materials Engineering, University of the Philippines, 1101, Diliman, Quezon City, Philippines

*E-mail: mdlbalela@gmail.com

Abstract. Future electronics devices are not only smaller and thinner, but are also flexible, bendable and even wearable. This evolution in technology requires direct printing of patterns onto any substrate using conductive inks made of a dispersion of metallic nanoparticles. In this study, Cl- ions was used to induce spontaneous sintering of silver nanoparticles (Ag NPs). Ag NPs with an average diameter of 56 nm were synthesized by polyol method using silver nitrate (AgNO3) and ethylene glycol (EG) as precursor and solvent, respectively. Poly(vinyl pyrrolidone) was used as the capping agent. Water-based inks were formulated containing different Ag NP loading (10–25 wt %). Using 50 mM NaCl aqueous solution as the dispersing medium, an ink with 15 wt % Ag exhibited a sheet resistance of about 2.85 Ω/sq. This very low sheet resistance was attributed to sintering of Ag NPs, which was accompanied by an increase in average diameter of nanoparticles from 56 to 569 nm.

1. Introduction

Conventional patterning of conductive line on a substrate is done through complex traditional steps, which may include electroplating, vacuum deposition, etching and photolithography. These processes are laborious, expensive, and energy intensive. Large quantities of waste are also produced due to corrosive chemical agents and inefficient materials usage [1-2]. Presently, researches are focused on conductive inks that can be printed directly onto thinner and more flexible substrates, i.e. polymer substrates, to create more complex circuit patterns [3-4].

Conductive inks are usually made of a dispersion of metal nanoparticles and a carrier solvent [3-4]. The small diameter of the nanoparticles minimizes nozzle clogging in print heads of industrial printers. However, the small particle size of nanoparticles induces agglomeration to reduce the surface energy [4-5]. Thus, stabilizing agents are added to prevent agglomeration and generally improve the dispersability of nanoparticles [4-8]. Examples of stabilizing agents include polyacrylic acid (PAA) [8-10], diethylamine [11] and gelatin [4-8].

Metal nanoparticle inks are typically composed of copper (Cu), gold (Au) and Ag nanoparticles. Among these metals, Cu is the most economical. However, it easily oxidizes, which significantly decreases its conductivity [1-2, 4-8]. Au nanoparticles exhibit high conductivity and thermal stability. But Au is too expensive for mass production [12]. Conductive patterns made by printing of metallic nanoparticles generally requires a post-treatment process, i.e sintering at high temperatures, to achieve the required conductivity. This is done to induce coalescence among the nanoparticles and ensure physical continuity for electron flow [11-12]. However, such harsh process complicates printing on heat-sensitive substrates, such as paper and plastics [3-4, 8-9].

There are different methods to achieve low temperature sintering, including low-pressure plasma [13], microwaves [14], laser [15], and use of chemical agents [3, 9, 16]. For example, patterns made of Ag NPs capped with PAA can undergo a built-in sintering mechanism due to exposure to NaCl [9]. Similar approach was used to induce sintering of Ag NPs capped with PVP [16]. In this study, sintering of Ag nanoparticle ink prepared with PVP was achieved at room temperature using NaCl as sintering agent. The effects of Ag loading on the physical continuity and electrical resistance of the pattern produced were also investigated.

2. Methodology

The Ag NPs were prepared by first dissolving 0.1 g AgNO3 and 0.4 g PVP (MW = 10,000) in 20 ml ethylene glycol, then heating at 160 °C for 14 min under reflux and constant stirring. The precipitates were then collected by centrifugation and washed several times with ethanol. The Ag nanoparticles were observed under scanning electron microscopy and analyzed by X-ray diffraction.

3. Results and discussion

Figure 1 shows an SEM image of Ag NPs synthesized at 160°C using polyol method with and without the sintering agent, NaCl. The Ag NPs have an average diameter of 56 ± 6 nm. The uniform size distribution is attributed to the similar growth rate of the Ag nanocrystals during synthesis, as well as to the steric hindrance provided by PVP. Then again, the presence of PVP layer around the nanoparticles prevents interconnection and typically results in high resistivity in printed conductive ink. The corresponding XRD pattern in Fig. 2 shows only peaks attributed to metallic Ag. Five distinct diffraction peaks at 2θ = 38.2, 44.4, 64.5, 77.5 and 81.6° were indexed to the (111), (200), (220), (311) and (222) reflections of metallic Ag (JCDPS File No. 04-0783). The sharp XRD peaks indicate the high crystalline structure of the as-prepared Ag NPs. No other peaks attributed to AgNO3 or other compounds were identified, which suggests high purity of the Ag product.

Fig. 1. SEM images of as-prepared Ag NPs (a) without and (b) with 50mM NaCl in the dispersing medium.

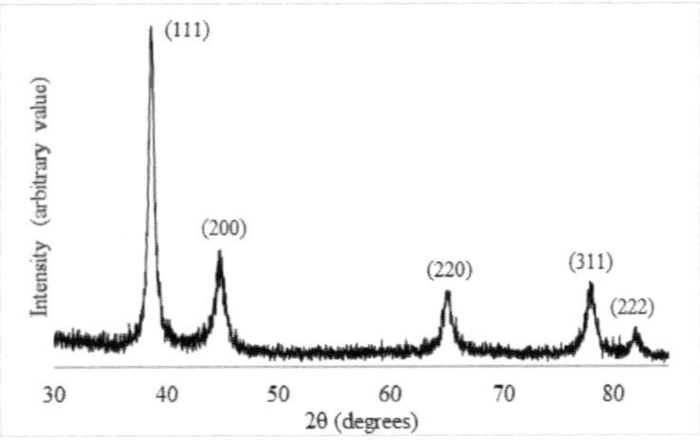

Fig. 2. Corresponding XRD pattern of as-prepared Ag nanoparticles.

On the other hand, an obvious increase in the particle size was observed after dispersion of Ag NPs in 50 mM NaCl aqueous solution. Large Ag particles with a mean diameter of about 569 nm were obtained as seen in Fig. 1b. This is approximately 916 % larger than the diameter of the as-prepared Ag NPs in Fig. 1a. It has been reported that Cl^- ions can cause the detachment of capping gents, such as PVP that surround the nanoparticles [3,9,16]. Due to high affinity of Ag with Cl^- ions, PVP molecures are possibly desorbed from the surface of the nanoparticles. The detachment of PVP probably causes destabilization, which increases the surface energy of the NPs. [3,9,16]. The atomic mobility caused by the increase in surface energy of the nanoparticles promotes agglomeration and their eventual sintering to further reduce the surface energy as shown in Fig. 3. The resistivity of 20 µL Ag NPs dispersed in 50 mM NaCl aqueous solution after deposition onto a glass substrate was measured to be about 2.85 Ω/sq. Without NaCl, no sheet resistance value was measured for the Ag nanoparticle ink at the same nanoparticle loading.

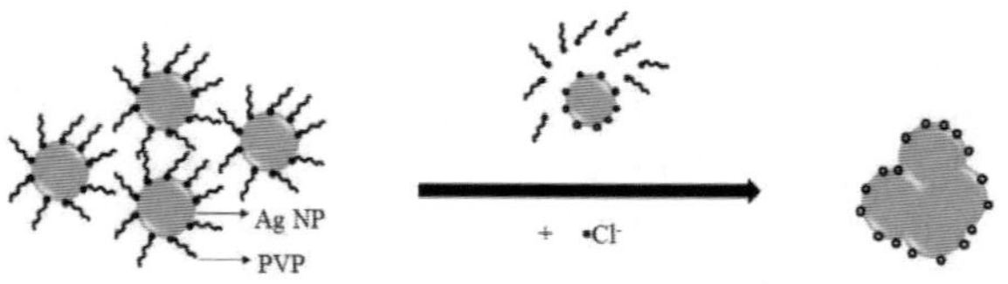

Fig. 3. Sintering mechanism for Ag NPs capped with PVP using Cl^- ions as sintering agent.

Instead of adding NaCl into the Ag NP ink, the effect of soaking in saturated NaCl solution of the Ag NPs pattern on glass substrate was also studied. Figure 4 shows the SEM image of a droplet of 20 wt % Ag NP in deionized water before and after soaking in saturated NaCl solution for 30 min. Similarly, the average diameter of the Ag NPs was increased to 303 nm, which is 441% larger than the diameter of the as-synthesized nanoparticles in Fig. 4a. The nanoparticles also have a wide size distribution, which could be attributed to Ostwald ripening [3,9,16]. The presence of large particles suggests that sintering and growth possibly occurred.

Fig. 4. SEM images of 20 wt % Ag NP in DI water (a) before and (b) after soaking in saturated NaCl solution for 30 min.

Figure 5 shows the change in sheet resistance with increasing AgNP loading in a water-based ink. Patterns were deposited on glass, followed by immersion in a saturated NaCl aqueous solution for 30 min. As seen in Fig. 5, there is a large difference between the sheet resistances of the ink containing 10 and 15 wt % Ag loading. At higher AgNP loading, more nanoparticles come into contact upon detachment of the PVP. The lowest sheet resistance of 97.33 mΩ/sq was obtained from the ink sample with 25 wt % Ag. In order to test the applicability of the developed Ag NP ink in paper electronics, the patterns were used to complete a circuit containing a power source and light-emitting diodes (LED). Essentially, all the patterns were able to light up the LED assembled in a circuit without any annealing as shown in Fig. 6. The paper also underwent bending and folding as shown in Fig. 6c.

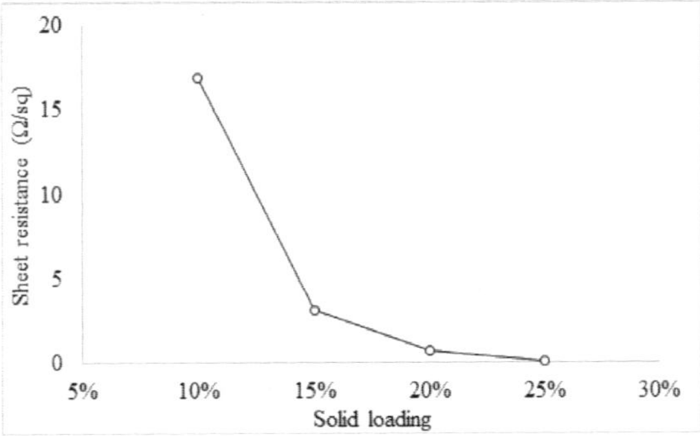

Fig. 5. A plot of sheet resistance (Ω/sq) with increasing Ag NP solid loading (wt %) for a water-based ink. Patterns were deposited on glass substrate then soaked in a saturated NaCl solution for 30 min.

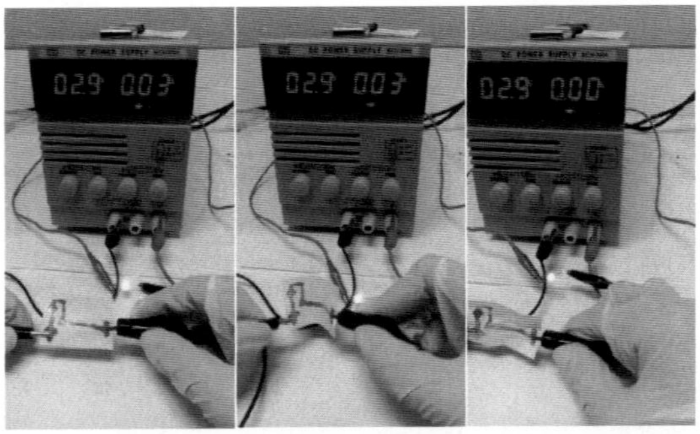

Fig. 6. A pattern made on a bond paper using the formulation containing 15 wt % Ag NP in 50mM NaCl solution being able to light up an LED (a) even after bending (b) and folding (c), the inset shows the bond paper bent in half four times.

4. Conclusion

Uniformly distributed Ag NPs with an average diameter of 56 nm were synthesized through a polyol method. PVP was employed as the capping agent. After dispersing the Ag NPs in 50 mM NaCl, the average diameter of the nanoparticles increased to 569 nm. The pattern achieved a sheet resistance of about 2.85 Ω/sq. Without NaCl, the pattern was not conductive. This indicates that Cl^- ions removed PVP on the surface of the nanoparticles, leading to their agglomeration and eventual sintering. The effect of Cl^- ions as a sintering agent was further explored by soaking the patterns made with increasing Ag loading in saturated NaCl solution. After 30 min soaking, the diameters of the nanoparticles increased, promoting true continuity in the pattern. As a result, the sheet resistance was decreased.

5. References

[1] Cui W, Lu W, Zhang Y, Lin G, Wei T and Jiang L 2010 *Colloids and Surfaces A: Physicochem. Eng. Aspects.* **358** 35-41.
[2] Ruvini D, Jha M, Amos D A and Druffel T 2013 *ACS Appl. Mater. Interf.* **5** 13227-13234.
[3] Magdassi S, Grouchko M, Berezin O and Kamyshny A 2010 *ACS Nano.* **4** 1943.
[4] Datu E M and Balela M D L 2016 *Key Eng. Mater.* **705** 163.
[5] Tan M, de Jesus L, Amores K L, Datu E and Balela M D 2014 *Adv. Mater. Res.* **1043** 114.
[6] Balela M D L and Amores K L 2016 *Adv. Mater. Res.* **1131** 255-259.
[7] Balela M D L and Amores K L 2015 *Sci. Diliman.* **27** 10-20.
[8] Wu C J, Chen S M, Sheng Y J and Tsao H K 2014 *J. Taiwan Inst. Chem. Eng.* **45** 2719-2724.
[9] Grouchko M, Kamyshny A, Mihailescu C F, Anghel D F and Magdassi S 2011 *ACS Nano.* **4**

3354.
[10] Huang Q, Shen W, Xu Q, Tan R and Song W 2014 *Mater. Lett.* **123** 124.
[11] Chen S P, Kao Z K, Lin J L and Liao Y C 2012 *ACS Appl. Mater. Interfaces.* **4** 7064-7068.
[12] Kang J S, Ryu J, Kim H S and Hahn H T 2011 *J. Elec. Mater.* **40** 2268.
[13] Reinhold I, Hendriks E, Eckardt R, Kraenburg J M, Perelaer J, Baumann R R and Schubert U S 2009 *J. Mater. Chem.* **19** 3384-3388.
[14] Perelaer J, de Gans B J and Schubert U S 2006 *Adv. Mater.* **18** 2101-2104.
[15] Ko S H, Pan H, Grigoropoulos C P, Luscombe C K, Frechet J M J and Poulikakos D 2007 *Appl. Phys. Lett.* **90** 141103.
[16] Tang Y, He W, Zhou G, Wang S, Yang X, Tao Z and Zhou J 2012 *Nanotech.* **23** 355304.

Acknowledgment
This study was supported by the Department of Science and Technology through the Philippine Council for Industry, Energy and Emerging Technology Research and Development under the research project: Up-Scale Synthesis of Metal Nanowires and their Application in Transparent Metal Nanowire Touch Panel.

IOP Conf. Series: Materials Science and Engineering **264** (2017) 012021 doi:10.1088/1757-899X/264/1/012021

Probabilistic analysis of the behavior of polymer matrix composite materials reinforced by different types of fibers

N. Harb [1, 2], B. Bezzazi [1, 2], S.Mehraz [1, 3], K. Hamitouche [1, 2], H. Dilmi [1, 2]

1 Research Unit Materials, Processes and Environment, University of Boumerdes, Algeria.
2 Coating Laboratory, Materials and Environment, University of Boumerdes, Algeria.
3 Research Center in Industrial Technologies CRTI, BP 64, Cheraga, Algiers 16033, Algeria.
E-mail: hnoura1542@gmail.com

Abstract. The requests of lightening of the structures and gains in performance lead to search for new materials and the associated processes for aeronautical and space applications. Long-fiber composites have been used for many years for these applications; they make it possible to reduce the mass of the structures because of their excellent compromise of mass/ rigidity / resistance. The materials in general contain defects which are essentially due to their nature and their mode of elaboration.
To this purpuse, we carried out a probabilistic analysis of the mechanical behavior in three-point bending of composite materials with a thermosetting matrix in order to highlight the influence of the number of folds of the fibers and the nature of the fibers on the dispersion of the defects in the stratified structures fiberglass, carbon fiber laminates and hybrid (carbon / glass) laminates.
From the results obtained, the dispersion of the defects is lower in the laminates of greater number of plies of the fibers and the hybrid laminates; the more the number of folds increases the more the mechanical characteristics increase; the hybrid laminates exhibit better mechanical properties compared to laminates of the same type of fiber. Finally, a morphological analysis of fracture structures and facies was investigated by scanning electron microscope (SEM) observations.

1. Introduction

The production of composite materials increases more and more, although their cost is higher than the traditional materials, they bring important advantages: Lightness, freedom of form, reduced maintenance, mechanical and chemical resistance. They make it possible to increase the life service of certain equipment due to their mechanical properties. Thanks to their chemical properties (resistance to corrosion), they reinforce safety due to their better resistance to shocks and to fire. They offer better thermal or sound insulation, and for some of them good electric insulation [1-3]. In composite materials with a polymeric matrix the fibers used as reinforcement are mainly synthetic fibers such as carbon or glass fibers, these fibers have very good mechanical and geometrical properties [4].
The study of the rupture behavior of composite materials with thermosetting matrix continues to be the subject of an abundant literature [5, 6].
The heterogeneous nature of these materials and the probabilistic aspect of the rupture are generally supplemented by rigorous statistical studies [7, 8]. The material architecture such as the orientation of

IOP Conf. Series: Materials Science and Engineering **264** (2017) 012021 doi:10.1088/1757-899X/264/1/012021

the reinforcements and the geometry create directional effects which influence the nature of the damage [9].
The objective of this work is in the context of a probabilistic characterization in static flexion of composite materials made up of a reinforced epoxy matrix with different plies and nature of the fibers in order to determine the weibull modulus which is a parameter material characteristic of the dispersion of the defects (or criterion of fracture sensitivity) for a given material.

2. Experimental procedures

2.1. Used materials
The materials used are composite materials with a polymer matrix, and reinforcements made of glass fibers and carbon with a fiber / matrix weight ratio of (3/1).

2.1.1. Epoxy resin: The epoxy resin is a thermosetting polymer, widely used in high performance composites in many industrial applications.
In this work we used MEDAPOXY 812 INJ from Granitex (ALGERIA). The mass ratio between the monomer and the hardener supplied by Granitex is MR = 2.
The characteristics of the MEDAPOXY 812 INJ resin are presented in Table 1.

Table 1. Characteristics of epoxy resin.

Weight Ratio (A / B)	2/1
Density (ISO 758)	1.1± 0.1
Viscosity (CF4/20°) (NFT30 014)	17sec (±2)
DPU : (NFP18 810)	30mn à 20°
Resistance to compression RC	> 70 (MPa)
Tensile strength Rt	> 57 (MPa)

2.1.2. Reinforcements: carbon fibers, glass fibers. The Carbon and glass fibers shown in (Figure 1) are synthetic fibers with very good mechanical and geometrical properties [4].
The properties of the reinforcements used are shown in Table 2.

Table 2. Properties of reinforcements used.

Reinforcement	Architecture	Surface mass(g/m2)	Density	Thickness (mm)	Tensile strength (MPa)	Elongation (%)
Carbon UD	unidirectional	230 ± 10	1.76	0.13	4300	1.8
Woven glass	Taffeta	225 ± 25	2.54	0.25	3100	4.4

Figure 1: a) fiber glass, b) carbon fiber. c) hybrid carbon/glass.

2.2. Method of preparation
The elaboration of materials is carried out by the vacuum method within laboratory of composite and plastic materials at the Algeria air maintenance base (Figure 2).

Figure 2. Vacuum *molding*

The composite materials studied in this work are laminated plates reinforced by different fiberglass and carbon plies,as well as hybrid laminated plates reinforced by alternating glass fiber/carbon plies of dimensions 300 x 300 mm,in the drying oven at 80°C during 8 hours after their production. Then the specimens are cut with a diamond disc and lubricated according to the AFNOR NF T 57-105 standard for mechanical testing[10,11].

2.3. Test device

The tests were carried out in three-point bending, carried out at ambient temperature on samples prepared according to the ISO 178 standard, which specifies the dimensions of the test piece (figure 3). The distance L between supports is selected according to the thickness (h) and the width (b) of the elaborated composite [4,12]. The test velocity is 2 mm / min. Tests are carried out on a machine of Zwick / Roell type equipped of the sensor of force of 2.5 kN and controlled by the TestExpert 9.0 software.

The results obtained show the evolution of displacement according of the load applied until the rupture. The bending σ_f stress, E_f Young modulus, the shear stress hear σ_{sh} are calculated respectively according to the following formulas:

$$\sigma_f = (3FL/2bh^2) \quad \text{[MPa]} \tag{1}$$

$$\sigma_{sh} = (3F/4bh) \quad \text{[MPa]} \tag{2}$$

$$E_f = (L^3/4bh^3m) \quad \text{[MPa]} \tag{3}$$

$$m = (f2 - f1) = \tan\alpha \quad \text{[MPa]} \quad \textit{(The slope of the elastic portion)}$$

σ_f and E are respectively the stress and bending modulus (MPa);σ_{sh} shear stress (MPa).
F: the maximum force (N). L: distance between supports (mm), L0: Length of the specimen (mm): ≥ 08 mm, b: Section width (mm); h: thickness of test piece (mm), f: the arrow of the specimen (mm).

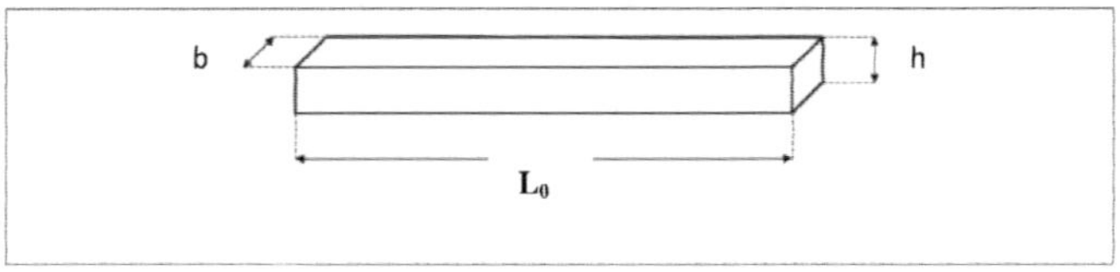

Figure3. Dimensions and shape of the test specimens for bending tests.

3. Experimental results and discussion

3.1. Force-displacement curves

Figure 4 shows examples of the evolution of force / displacement curves for the various laminates. The shape of the curves is generally similar for laminates of the same type of fibers.

IOP Conf. Series: Materials Science and Engineering **264** (2017) 012021 doi:10.1088/1757-899X/264/1/012021

These curves are generally linear until fracture reflecting the fragile character of glass fiber laminates. However, carbon fiber and hybrid laminates are distinguished after a large linear portion which represents the elastic response of the material by a second deviating part of the linearity. This reflects the triggering and accumulation of damage that takes place within the structure before the total collapse. This diffuse and progressive damage is mainly caused by a multi-cracking of the matrix, A mechanism for loosening and detaching the fibers, Fiber-matrix decohesions which limit or prevent stress carry-overs between the broken fibers and the intact fibers as well as individual ruptures of the fibers which are minor.

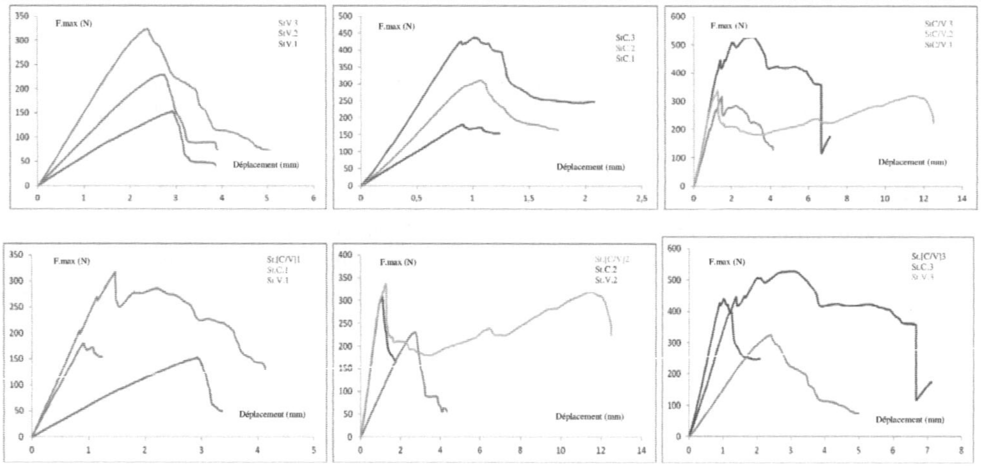

Figure 4: **I.** Influence of number of folds of fibers on the flexural behavior of laminates: -a) glass fiber, b) carbon fibers, c) carbon / glass hybrid. **II.** Influence of the nature of the fibers on the flexural behavior of the laminates: d) has 8 plies, e) has 10 plies, f) has 12 plies.

A significant variation in flexural behavior of the various laminates is observed depending on the number of fiber folds and the nature of the fibers.

We can say that the more the number of folds increases from 8 to 12 plies, the more the modulus of elasticity in inflexion increases from 23.97 to 29.74 GPa for the fiberglass laminates. From 55.90 to 61.99 GPa for the fiber laminates of Carbon. From 56.78 to 60.92 GPa for hybrid laminates. The bending stress increases from 236.10 to 286.26 MPa for glass fiber laminates. From 330.20 to 411.50 MPa for carbon fiber laminates. From 423.22 to 465.33 MPa for hybrid laminates. Shear stress increased from 5.33 to 8.35 MPa for glass fiber laminates, from 7.13 to 1080 MPa for carbon fiber laminates and from 9.80 to 12.20 MPa for hybrid laminates.

Table 3 summarizes the mean values of the measured mechanical characteristics of the various laminates.

It is also noted that hybrid laminates exhibit better mechanical characteristics except for the 12- plies hybrid laminate which presents a decrease in young's modulus compared to the 12-folds carbon fiber laminate from 61.99 to 60.92 GPa.

Table 3. Mechanical characteristics of the various laminates.

Sample	thickness h(mm)	Section width b(mm)	F.max (N)	$f_{F.max}$ (mm)	σ a bending (MPa)	Shear stress σ_{sh} (MPa))	Young's modulus E (GPa)
Stv1	2.10	10.70	151.07	2.81	236.10	5.04	23.97
Stv2	2.29	10.42	222.57	2.70	278.38	6.99	26.99
Stv3	2.83	10.29	287.67	2.34	286.26	7.40	29.74
Stc1	1.90	10.90	188.77	0.97	330.20	6.83	55.90
Stc2	2.23	10.85	295.33	1.11	390.96	9.15	59.15
Stc3	2.70	10.95	458.72	1.04	411.50	11.63	61.99
St.[c/v]1	2.01	10.95	323.24	1.57	423.22	11.01	56.78
St.[c/v]2	2.35	10.25	332.56	3.18	439.24	10.85	59.95
St.[c/v]3	3.36	10.30	520.77	2.98	465.33	11.29	60.92

St.V: fiberglass laminate. **St.C**: carbon fiber laminate. **St.[C/V]:** Hybrid (carbon / glass fiber) laminate.
(1) Corresponds to 8 folds. **(2)** Corresponds to 10 folds. **(3)** Corresponds to 12 folds.

4. Probabilistic approach: probabilistic model of wei bull

The material rupture is controlled by the distribution of defects in volume related to the production process and to the characteristic values of the materials used…ect, this set of parameters is randomness..

According to Weibull, the probability of rupture P_f of a material of volume V subjected to a uniform distribution of stresses σ, is given by [13,14]:

$$P_f = 1 - exp\left[-V\left[\frac{\sigma-\sigma_u}{\sigma_0}\right]^m\right] \tag{4}$$

σ_u and σ_0 are respectively the threshold stress below which the probability of rupture is zero (consider $\sigma_u = 0$) and the standardization constant. The most technique used for determining the weibull parameter is the linearization of this equation, which allows us to write:

$$Ln\left[Ln\left(\frac{1}{1-P_f}\right)\right] = m.Ln(\sigma - \sigma_u) + Ln(V) - mLn(\sigma_0) \tag{5}$$

Representation of $Ln\left[Ln\left(\frac{1}{1-P_f}\right)\right]$ in terms of $Ln(\sigma - \sigma_u)$ is a line of slope m.

The probability of rupture is calculated from the expression:

$$P_f = \frac{i}{N+1} \tag{6}$$

It is a question of assigning a probability of rupture to each stress level as a function of rank i and according toan ascending order.

4.1. Analyse probabiliste

The probabilistic aspect of the rupture was highlighted by an analysis of the results from the application of the weibull statistical model [7]. Figure 5, Shows the variation of the rupture probability according to the evolution of the bending stress in flexion.

IOP Conf. Series: Materials Science and Engineering **264** (2017) 012021 doi:10.1088/1757-899X/264/1/012021

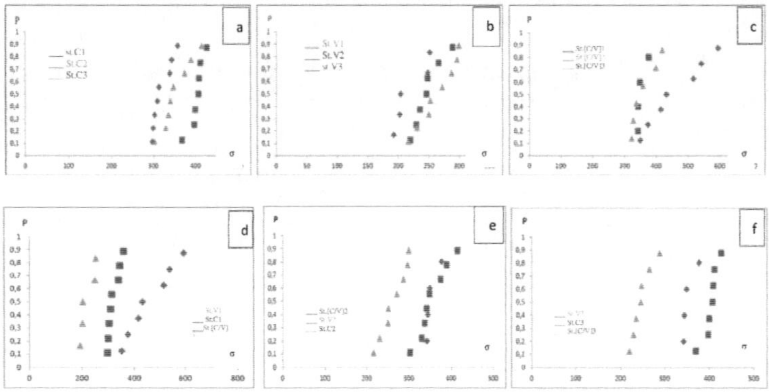

Figure 5. Probability variation Pf as a function of the stress at break σ: a) laminate a 8,10 and 12 plies of carbon fiber. b) laminate has 8,10 and 12 plies of fiberglass. c) 8,10 and 12 ply hybrid (carbon / glass) laminate. d) laminates with 8 plies of fiberglass, carbon and hybrid (c / v). e) laminates with 10 plies of fiberglass, carbon and hybrid (c / v). f) laminates with 12 plies of glass fiber, carbon and hybrid (c / v).

- Weibull modulus calculation for each type of laminate:

Examples of plots determining the weibull modulus in the case of stress at rupture are illustrated in Figure 6.

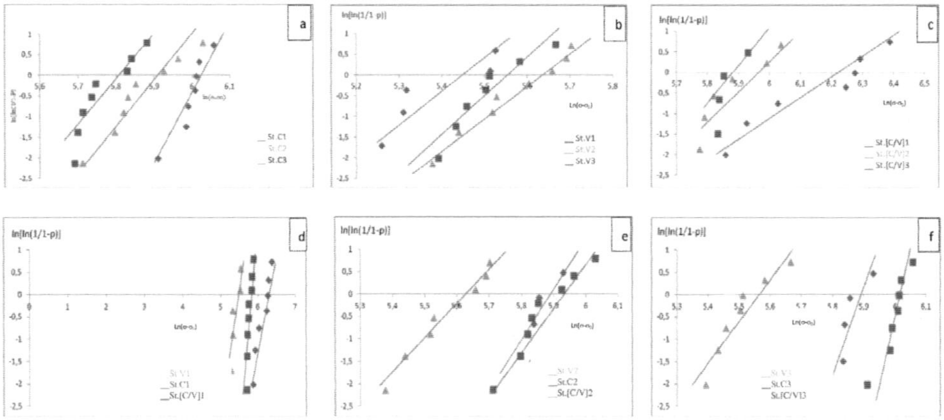

Figure 6. Variation of the weibull modulus according to the number of folds: a) laminates with 8,10 and 12 plies of carbon fiber. b) laminates with 8,10 and 12 plies of fiber glass. c) hybrid laminates with 8,10 and 12 plies of fiber (C / V) and depending on the nature of the fiber. d) laminates with 8 plies of glass fiber, carbon and hybrid (c / v). e) laminates with 10 plies of fiber Glass, carbon and hybrids (c / v). f) laminates with 12 plies of glass fiber, carbon and hybrid (c / v).

Table 4 summarizes the average weibull modulus values obtained for different laminates.

Table 4. The average weibull modulus values.

laminate	**St.V1**	**St.V2**	**St.V3**	**St.C1**	**St.C2**	**St.C3**	**St.C/V1**	**St.C/V2**	**St.C/V3**
Weibull modulus (m)	6.90	7.06	7.18	8.50	8.72	11.56	7.36	9.38	12.59

These weibull modulus values are included in the interval relative to composite materials [15,16]. A slight variation in weibull modulus of 6.90 to 7.18 was observed depending on the number of folds of the glass fibers laminates. A significant variation from 8.50 to 11.56 for the carbon fiber laminates and from 7.36 to 12.59 for the hybrid laminates.We notice that Weibull modulus vary according to the fiber's nature and the number of plies.

5. Morphological analysis

A morphological analysis of fracture facies and structures of different materials was investigated by scanning electron microscope (SEM) observations and presented in Figures 7 and 8.

Figure 7. Facies of the rupture of laminates: 1) St.V, 2) St.C, 3) St.C/V.

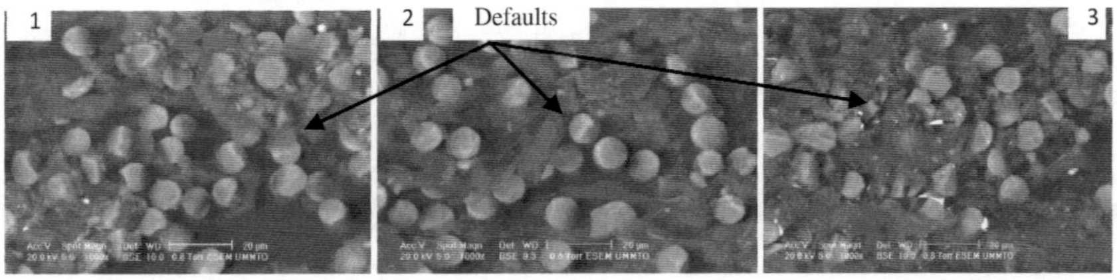

Figure 8. morphology of stratified structures : 1) St.V, 2) St.C, 3) St.C/V.

The crack generally follows a perpendicular plane to the load, glass and carbon fiber loosening, fiber breaks, delamination and cracks in the matrix have been observed.

Laminated composite materials contain defects which are essentially due to the nature of the fibers and the number of fibers used.

6. Conclusion

The influence of the nature of the fibers and the number of fibers on the mechanical behavior and defects in composite structures was determined by the weibull probabilistic analysis of the flexural behavior. The experiments are carried out on laminated composites with an epoxy matrix and differentiated by their natures of fibers (glass, carbon and hybrid[c/v]) and by their numbers of plies (8plies, 10plies and12 plies).

The analysis of the results of flexural behavior of the various laminates leads to that conclusion:

- More the number of folds increases the higher the mechanical characteristics,
- Hybrid laminates have better mechanical properties than laminates of the same type of fiber, except for the 12-plies hybrid laminate which has a decrease in young's modulus compared to the 12-plies carbon fiber laminate.
- In hybrid laminates and laminates with greater number of folds of the fibers, the dispersion of the defects is weaker.

The use of weibull statistics allowed us to calculate the weibull parameter for the case of the flexural rupture stress of the laminates with different plies and the nature of the fibers.

The Weibull modulus, m, is a material parameter characteristic of the dispersion of defects (or criterion of fracture sensitivity) for a given material.

- When the Weibull module is weak, there is a large dispersion of defects in the material. Consequently, there will be a great dispersion on the values of stresses of activation of the defects within the volume v of the material studied.
- When the Weibull modulus is high, the dispersion of defects is small, as well as the dispersion on the stresses of activation of the defects within the volume v of the material studied.

7. References

[1] Daniel Gay, " Matériaux Composites ", Ed. Hermès, ISBN : 2 86601 586 X. paris, 1997.
[2] Reyne M. Technologie des composites. Edition Hermès 1998.
[3] Reyne M. ''l'utilisation des matériaux de synthèse dans les moyens de transport''. Matériaux et techniques n 7-8, pp3-8, 1991.
[4] Dixneuf E., Lamy B.et Bruno M.A., 98 '' Spécificités du comportement en flexion 3 trois points des matériaux monolithiques et sandwichs à renforts tissés, non tissés et tricotés '',Compte rendus de JNC 11, V.1,Arcachan,pp.99-107.
[5] Chafra et Chevalier, 01,''Endommagement orienté et rupture des matériaux composites sous chargement cyclique''XV congrès français de mécanique, 3-7 Sept., Nancy.
[6] Berthelot. J.M., " Matériaux composites, comportement mécanique et analyse des structures, Ed. Lavoisier, 2005, ISBN : 2-7430-0771-0, pp 172.
[7] Redjel et Decharentenay,87,''Application des statistique de weibull à la caractérisation des résines phénoliques et des matériaux composites SMC'', Matériaux et Technique,N 10-11, pp.421-424.
[8] Guillaumat L. et Lamon J.94,''Simulation statistique – probabiliste du comportement mécanique non- linéaire d'un composite 2D SiC/SiC'', comptes rendus de JNC9, V.2,22-24 Nov.,Saint-Etienne, 1063-1072.
[9] Chafra et chevalier, 01,''Endommagement orienté et rupture des matériaux composites sous chargement cyclique'', XV congrès français de mécanique, 3-7 Sept., Nancy.
[10] D. Harper and M. Wolcott. Comp. Part A: Applied Sci. and Manuf. 35, (2004), 385-394.
[11] Aranberri, T. Lampke and A. Bismarck. J. of Colloid and Interf. Sci.2003, 263 (580-589)
[12] R. G. Raj, B. V. Kokta and C. Daneault. J. of Materials Science, 1990, 25, 1851-1855.
[13] Benamira M., 03''caractérisation à la rupture d'un matériau composite utilisé par L'ONAAPH'', these de magister, Université de Annaba, Algérie.
[14] Ardillon E. and Verite B. ''Example d'évaluation probabiliste de coefficients de sécurité dans le cadre du RSE-M''. Rapport technique n. HP-16/96/031/B, EDF-DER, 1996.
[15] Zhao.F.M.& Takeda.N.,''Effet of interfacial adhesion and statistical fiber strength on tensile strength of unidirectional glass fiber/epoxy composite.Part.II.comparaison with prediction composites Part A, Vol.31, 2000, pp.1215-1224.
[16] Gao.S.L., Mader.E, Abdkader.A.&Offermann.P., ''Environmental resistance and mécanical of alkali-resistant glass fibers with surface Sizengs'', journal of non cristalline solids'' V.325, 2003, pp.230-241.

Institute of Physics
Dirac House, Temple Back
Bristol BS1 6BE UK

ISSN: 1757-8981
ISBN 978-1-5108-5310-2

CURRAN ASSOCIATES INC.
proceedings.com

4th International Conference on Advanced Composites and Materials Technologies for Arduous Applications (ACMTAA 2015)

IOP Conference Series: Materials Science and Engineering
Volume 153

Wrexham, United Kingdom
5 – 6 November 2015

Institute of Physics